THE CALCIUM CONNECTION

THE CALCIUM CONNECTION

The Little-Known Enzyme at the Root of Your Cellular Health

BRUNDE BROADY

with Karen Lacey and Lyle Wilson

FOREWORD BY RUSSELL DAHL, PhD

SKYHORSE PUBLISHING, INC.

New York, New York

Skyhorse Publishing books may be purchased in bulk at special discounts for sales promotion, corporate gifts, fund-raising, or educational purposes. Special editions can also be created to specifications. For details, contact the Special Sales Department, Skyhorse Publishing, 307 West 36th Street, 11th Floor, New York, NY 10018 or info@skyhorsepublishing.com.

Visit our website at www.skyhorsepublishing.com.

10 9 8 7 6 5 4 3 2 1

Library of Congress Cataloging-in-Publication Data is available on file.

Book and cover design by Charles McStravick
Cover photo by Gonzaga Gómez-Cortázar Romero

Print ISBN: 978-1-5107-6391-3
Ebook ISBN: 978-1-5107-6392-0

Printed in the United States of America

To Robert and Knute

CONTENTS

SECTION FOUR
THE CALCIUM ATPASE MAX PROTOCOL (CAMP)

FOREWORD

by Russell Dahl, PhD

As scientists, we spend a considerable amount of energy in the search for singular theories or hypotheses that can explain multiple observed phenomena. What we desire are logical concepts that have the ability to tie together many seemingly disparate observations in a tidy package. We can think of this as a unified theory. The universal governing rules that are uncovered in these discoveries have led to some of the most significant milestones in our scientific understanding.

Consider for a moment the huge leaps in scientific understanding that we have gained from some of these unifying theories. For instance, the theory of evolution gave us answers for such fundamental questions as causes of species extinction, origins of biodiversity, and how organisms adapt, all while providing a framework for the origin of life and how our Earth came to be populated. Likewise, gene theory gave us a toolbox with which to explain heredity, traits, and a means for rationalizing the origins of many diseases. Quantum mechanics gave us rules-based methods for understanding the nature of light, colors, shadows, and, ultimately, the composition of matter itself. The medical research community is on a constant search for unifying theories that can be applied to the causes of disease. In this book, Brunde Broady lays the groundwork for a compelling theory about the causes of disease, while providing ample evidence from the primary scientific literature that Calcium ATPase is a major determinant of human health and wellness.

I first met Brunde in the context of my work studying relationships between intracellular calcium disturbances and disease. Early on in our discussions, I recognized a mutual appreciation for the potential of calcium handling in human health and disease pathogenesis. We both share a common familiarity with the seminal studies in the field that have provided compelling evidence for the importance of Calcium ATPase. I was immediately impressed by her voluminous and almost encyclopedic knowledge of Calcium ATPase, specifically in its relationship with health matters. She has the unique ability to have high-level scientific discussions in one instance, and then easily shift to convey these complex concepts to the lay public. This contributes to the importance that this book will have in raising the visibility of this important area.

I have spent the majority of my career studying the various proteins responsible for calcium ion handling in our cells, with a focus on figuring out how to modulate them with small molecules. The ultimate goal is to develop therapies for various diseases. I have focused on a specific calcium pump, called the Sarco/Endoplasmic Reticulum Calcium ATPase, SERCA for short. I worked on this protein first as the leader of small molecule research at Celladon, where our major clinical program focused on a gene therapy product that actually delivered SERCA to failing human hearts. I subsequently founded a company, Neurodon, to continue our important work on developing therapeutics that modulate this important enzyme fundamental to intracellular calcium homeostasis.

Over the last decade, we have established a pipeline of molecules that target the activation of SERCA. Using these compounds, we have been able to establish critical links between compromised or insufficient Calcium ATPase activity and diseases as diverse as Alzheimer's disease, diabetes, Parkinson's disease, muscular dystrophy, and others. We have demonstrated that the activation of Calcium ATPase in animal models of these diseases can reverse or halt disease progression, and we have long realized that Calcium ATPase is a special biological target with immense potential for therapeutic intervention.

Needless to say, I was elated to read this book by a fellow Calcium ATPase evangelist, and I am confident that Brunde's book will be integral in helping to publicize the importance of this enzyme and how it can help patients and the public at large. Our discussions and the insight contained in this work have led to many productive initiatives that have served to push forward my lab's research in this area.

A key takeaway of this book is that Calcium ATPase is an exquisite barometer of human health. This represents an example of the grand unifying concepts that I alluded to earlier and, as such, has the potential to greatly advance our understanding of human health and disease. What is also significant is that Brunde has been able to convincingly present Calcium ATPase function in a manner that highlights its potential to maintain or restore health. This extensively researched volume provides readers a detailed compendium of information. Our findings in the lab completely back up the notion that Calcium ATPase is a linchpin of cellular health.

The fact that the basis for this work is the primary, peer-reviewed literature from prominent scientists in the field makes this book stand out from many of the health and wellness books available today.

As a scientist with over two decades of independent research experience, I have found *The Calcium Connection* to be valuable in advancing my own knowledge of Calcium ATPase outside of my particular area of expertise. I look forward to this book inspiring a revolution in terms of our thinking about the origins of disease, and I am excited with the prospect that this important work will help the public at large be aware of this important biological target.

Russell Dahl, PhD
CEO and scientific founder of Neurodon
Crown Point, Indiana

PREFACE

Brunde's Story

The birth of my understanding of Calcium ATPase began with another birth, one packed with the depth of love only a mother can know. Hours after my son, Knute, was born, I was in my hospital room breastfeeding when the nurse came in. Knute had fallen asleep, as peaceful as an angel. But as the nurse adjusted the blankets, a look of concern crossed her face.

"He's fallen asleep," she said, as if I'd done something wrong.

"I know." How could anything be sweeter, more perfect, than my newborn son's sleeping face? His long, still eyelashes, his miniature fingers that lay lightly curled in a tiny fist.

"He's turning blue," she said and lifted him from my arms. "He's not breathing right."

She whisked away my son, who was only hours old, into the ICU, where I couldn't follow. I was horrified, panic-stricken. Was he okay? Would he die? What were they doing to him? But I lay trapped in my bed, still recovering from childbirth.

Knute was kept in ICU that first week. The doctors ran scores of tests, but the only diagnosis they could come up with was sleep apnea. That explained why he stopped breathing while snuggled in my arms fast asleep. The solution was for him to wear an apnea monitor full-time, every day, every minute. When it let loose its high-pitched beep, we had to check and make sure he was still breathing. So in effect, he was never left alone for the first three years of his life. Someone, usually me, was with him every second.

At the time, I was staying with my parents in Texas while my husband looked after things at our home in New York City. I hired an amazing caregiver named Dana who immediately bonded with Knute, becoming as emotionally invested as if he were her own. I don't know what I would have done without her in those early months. Knute was with one or the other of us every minute. When one of us slept, the other one was with him. The two of us became fine-tuned to every subtle nuance of his breathing.

While we got the sleeping and monitoring under control (minus perpetual exhaustion), his feeding was difficult. He did not swallow properly and aspirated his food, which led to pneumonia. His inflamed and narrowed airway gave way to stridor, the high-pitched wheeze of his struggling infant lungs.

At four months, Knute had a swallow test. He'd been slobbering a lot, and as a new mother, well, I didn't know what was happening. The test was to see if his milk was going down the esophagus or, more dangerously, into the lungs. The news wasn't good. I now needed to thicken his milk with rice cereal so it would be solid enough for Knute to swallow without going into his lungs. We also had to start administering four nebulizer treatments each day just to control the fine balance of eating and breathing.

Throughout Knute's treatment, seeing my sweet baby boy going through all these procedures tore me up. He was infinitely vulnerable, and I was infinitely responsible for his care. I sat quietly watching him, aware of every breath, every heartbeat, knowing I would do anything to make him safe.

Over time, with the guidance of who I hoped were wise doctors, Knute stabilized. His apnea remained under control, and although we designed our lives around his sleep schedule and the apnea monitor, he ate well and grew on schedule.

Sort of.

He was always . . . droopy. It was hard to tell at that age, because babies are babies. But as he grew, he didn't develop the core body strength he needed to hold himself upright in his stroller or the jogger—he would just droop. As a result, I carried him everywhere.

At about six months, his eyelids drooped, too, especially on one side. The doctors told me it was *ptosis* (neuromuscular weakness of the muscles that control the eyelids), and although they weren't concerned, I was.

Given his persistent eyelid ptosis and overall lack of core strength, he was tested for myasthenia gravis (a rare but long-term neuromuscular disease). Negative.

Knute hit his developmental milestones, but always at the end of the normal period. Again, the doctors weren't worried. They had already taken care of the potentially life-threatening issues of sleep apnea and swallowing. "Let's just wait and see," they told me. "All kids progress at different rates."

But I was his mother, and I knew something wasn't right.

I watched him carefully. I noticed how he reacted to what I fed him. When he was around nine months, I started expanding his diet with fruit puree, yogurt, Cheerios, and other typical starter foods. Knute began to develop horrible little nodules all over his body, like mosquito bites. I took pictures, because I wanted to be able to show the doctors and maybe other people within the medical community. One doctor suggested it could be mites, so Knute received several mite treatments that consisted of covering his body in a lotion for several days—a lotion that I was unaware contained toxic pesticides.

We went through allergy testing, and everything came up negative. Knute wasn't allergic to anything, yet after eating certain foods he broke out in nodules. And not just that, his mood changed as well, in ways only a mother can see. His asthma was triggered, his droopiness increased. I could see the connection, but why couldn't anybody else? None of the doctors could diagnose what was happening or knew what to do about it. My sense of desperation deepened.

At this point, Knute was a year and a half old, and he'd had symptoms his entire life. Yet the medical community, for all their hard work and good intentions, had come up with no solid determination for what was wrong. Yes, we'd taken care of the most severe symptoms, but nothing was diagnosed, and most certainly nothing was cured.

All this was so tangible for me; Knute was right there, and I knew how he was reacting to certain foods. I began to do my own research. I began to document everything he ate and his reactions. I knew it wasn't myasthenia, but I also knew his neuromuscular symptoms were similar. From there I researched rarer, congenital versions of myasthenia. The doctors I worked with didn't even know these versions of the disease existed. His symptoms could have been signaling slow-channel syndrome, caused by too much calcium getting in the cells or perhaps it could be some type of metabolic disease, but tests turned up negative. Nothing linked up completely with all the various symptoms. I researched myasthenia and understood it was related to muscle contraction. I understood that calcium was a key component in the contraction and relaxation of muscles, but what about all the other symptoms—asthma, hives, mood swings? With little help from the medical community, I had no solutions at hand. Nothing I could grasp onto for a cure or to even mitigate my son's symptoms.

Our home in New York City is a one-bedroom apartment, twenty-one floors up, with a sprawling veranda. While there was no room inside the apartment for my ever-increasing stack of research papers and files, we did have a garden shed on the veranda. My husband and I cleared room in the shed for a desk, and I set up my office. In the winter I was freezing, and in the summer I sweated and swatted at flies. But I had a space to clear my head, organize my thoughts, and dig deeper into my son's health mystery.

I continued to try and understand what was triggering his symptoms. Through Google searches, I found a group of concerned parents both in the United States (the Feingold Association) and in Australia (the Food Intolerance Network) who had made a connection between their children's behavior and food additives, specifically food dyes and additives. Several of the symptoms referenced were the same as Knute's, but I was still stumped over why. Why were asthma and hives and ptosis all present? What was the underlying mechanism? Myasthenia and slow-channel both pointed to calcium, so I thought, why not? It's as good a tip to follow as any.

The Australian group had been interested in BHA and BHT, two ubiquitous food additives. On a hunch, I wondered if these could be related to intracellular calcium. I searched on PubMed and learned that these additives had a negative impact Calcium ATPase, a fundamental regulator of intracellular calcium. My first real clue was that everything Knute reacted to was a Calcium ATPase inhibitor.

Progress.

But if calcium is related to muscles, then how did the allergic reactions fit in? And mood changes weren't muscular. I could see the pieces, but no matter which way I turned them, they wouldn't click together.

That is, until I understood mast cells, a type of white blood cell. Reduced Calcium ATPase levels in these cells trigger mast cell degranulation, prompting the release of histamine and leukotrienes—the inflammatory factors that cause the allergic response of welts and asthma.

Another piece clicked together.

I then backed into an understanding that intracellular calcium levels in neuron cells determine the release of neurotransmitters. Neurotransmitters are chemicals in the brain, such as dopamine, serotonin, and acetylcholine, that affect a wide range of physiological processes including digestion and behavior. This could explain my observations regarding Knute's mood changes.

Despite their superficial diversity, every one of Knute's symptoms linked back to intracellular calcium levels. Maybe I wasn't insane after all.

All kinds of factors contribute to inhibiting Calcium ATPase: natural, seemingly healthy foods (such as organic cookies with aluminum-based baking powder), a plethora of food additives and dyes, pesticides, off gasses, and many more. I thought back over the last couple of years and realized that I hadn't paid any mind to those kinds of things. I hadn't been providing organic, additive-free food. We had a flea problem that we treated with pesticide bombs. We had our wooden floors refinished, which gave off fumes for weeks. Bathrooms were cleaned with bleach-based cleaners, and our older house did not have much ventilation. From there, I decided to cut back on everything that inhibited the enzyme Calcium ATPase (which I go into in great detail in Section

Three). I couldn't test his levels of Calcium ATPase, but I could assess them based on his symptoms.

When I cut out exposure to the inhibitors, he got better. It was as simple, and complex, as that. Though now everyone thought I was nuts again, because I was so adamant about what he could and couldn't eat. My well-meaning sister would give him colored popsicles along with her children, and my parents could not resist stopping at McDonalds for chicken nuggets (with tons of chemical additives) or a sausage biscuit (the worst on any level). It took many such instances to convince them of the connection between food bumps, asthma, and, most disturbingly, his mood. Finally, they got it. It was quite a job packing simple foods for our annual trip to the beach on a small island that had no organic food. I would take a cooler of clean foods and cook for him. It was not easy. A little boy wants to eat all of the same hotdogs, ice cream treats, and cotton candy as everyone else. School lunches provided another challenge. Knute was the only child in the school who was allowed to bring his own lunch. (Thank goodness the school allowed this, even though it was not due to typical food allergies such as peanuts.) In all of these situations, I eventually learned that from time to time he could splurge on an "off" food. These instances took some pressure off, and that helped.

Today, I believe that Knute's body had a tough time with calcium regulation, so any outside factor that inhibited his already weakened facility resulted in symptoms. As you will soon learn, calcium regulation is complicated. Knute's sensitivity to Calcium ATPase inhibitors does not directly indicate that he has reduced Calcium ATPase levels. (In fact, there are specific genetic diseases such as Darier's disease, which is related to reduced level Calcium ATPase in the skin, as well as rare muscle diseases. However, these diseases do not manifest in the wide range of Knute's symptoms.) I may not have a definitive answer in terms of what caused his sensitivity to Calcium ATPase inhibitors, and quite frankly I do not care. What mattered to me (and what still does) is that I was able to significantly improve his health and thus his quality of life. And by the age of five, my sweet son was conversant in intracellular calcium regulation!

Although I had gained an understanding and awareness of Knute's condition, attending doctors and other health professionals just stared back glassy-eyed when I told them about my discoveries. So I gave up on sharing my findings. Instead, I kept learning. The papers in my garden shed piled higher as I realized that intracellular calcium regulation was a primary factor in heart disease, obesity, prenatal and early childhood development, and even in the rate of recovery after athletic exertion.

Today, Knute is a healthy, thriving young man. His health problems are largely behind him, and he enthusiastically pursues his academic passions through specialized science research programs and is majoring in mechanical engineering at a top academic institution. He has come to self-modulate his food choices and has learned from experience to avoid his "off foods," especially during particularly important periods such as exam time, when he diligently sticks to a completely clean diet. He cannot afford to be off. It affects his mood, his performance, and his skin.

Knute will never be an athlete, but his early childhood experiences have left him with a depth of empathy that continues to astound me. He knows things go wrong in life, and that not everyone is born with the same set of skills; that you certainly can't blame someone for not trying hard enough if their physiology simply doesn't allow for certain outcomes, whether it be playing on the school football team or chess club. He's able to see other people with a clarity and compassion I can only one day hope to achieve. This entire journey, and the help I hope to provide other people, is solely because of him.

AN OVERVIEW OF WHAT'S AHEAD

- The first section lays a foundation for the rest of the book by providing a basic understanding of what Calcium ATPase is, and how it works.
- The second section contains a considerable amount of science concerning the role of Calcium ATPase in the body, specifically on how it relates to numerous chronic diseases, such as cancer, heart disease, Alzheimer's, and diabetes.

You do not have to read about every disease state to benefit from this book. Feel free to focus on the concerns that matter to you. I have also included a summary at the end of each chapter—there is a lot of new information to process!

- The third section outlines toxins that inhibit Calcium ATPase. It's quite information-dense; I felt committed to include this detail in order to make a definite and indisputable case for the impact our environment has on Calcium ATPase. However, if you are pressed for time or feel like you already know the fundamentals about toxins, I have included a chart at the beginning of each chapter that summarizes each toxin and its relationship to Calcium ATPase.
- The fourth section details the Calcium ATPase Max Plan (CAMP) protocol. This is where things get exciting! I will lay out concrete actions you can take to support your Calcium ATPase levels. It includes a special section written specifically for parents and provides guidance to how best to navigate a child's Calcium ATPase levels.

ACKNOWLEDGMENTS

Most importantly, I would like to thank my coauthors Lyle Wilson and Karen Lacey. This book would not have happened without their support, talent, encouragement, and perseverance. Lyle is a gifted health practitioner and educator, and Karen is both a novelist and a bestselling author. Their combined efforts added A LOT to every page of this book! Their integrity of mission matched that of my own and helped me keep my eye on the goal of getting this information out to help the world.

I want to thank culinary expert Stefanie Sacks who stepped in at the right moment to author the CAMP recipe section. Stefanie also made valuable contributions to the other CAMP program chapters in both substance and organization. Her quick mind and quick eye were priceless.

I would like to wholeheartedly thank my marketing team at Pivotal Twist. Sean Perlmutter and Henry Caplan provided their experience, talent, and support above and beyond what I could have hoped for. Their talented team, including Indigo, Megan, Nicole, and Zia, helped me with all content needs in a creative and thoughtful way. (And thank you, Devon Nola, for introducing me to Henry and Sean!) I also want to thank Carrie Simons and her team at Triple 7 Public Relations for putting all hands on deck to do an awesome job getting the much-needed PR for my book.

Thanks also to Kevin Anderson, founder of Kevin Anderson and Associates, for providing crucial support in all aspects of the book-writing and publishing processes (including introducing me to Karen!). From the very start, Kevin believed in my book and provided the resources I needed to help me realize my vision. Thank you, Hector Carosso, at Skyhorse for allowing me to stay in control of my process and twisting arms to get me the extra time I needed those last few weeks! Thank you also for the team at Target Marketing Digital for helping me define my message for the online world from scratch.

Thanks to Charles McStravick for doing such an incredible job designing my book cover and interior pages. No one could have done better! Thanks, Kate Hanley, at KAA for your editorial direction. Thanks to Meg Descamp for line editing and Maria Alcoke for images.

Thanks to my brother John, who was with me every step of the way—reading outlines, drafts, encouraging me to continue. I never could have done it without you, Johnnie. My sister Lynn, who generously opened her home, printer, and even kitchen table to my mad frenzy of research in the early days. My brother Vincent, who believed in me before most. My Mom, whose natural curiosity and bent towards science inspired me to search for answers. To my father for being a pioneer, and always believing in forging your own path through difficult times and never giving up. For my Aunt Judith, who encouraged me with boundless belief.

Thanks also to my dear friends Lavina, Jody, and Nagla. I am so fortunate to have stumbled upon the miracle of your friendship and love. Also to Michael Porder, who drilled into me Rome was not built in a day (and gave me much love and support). To Nancy Austin for her faith in the foundation of my work. To Yoav Cohen for being the lighthouse during the storm when I was lost at sea.

Thank you to John Robbins, Dr. Dana Cohen, Linda Bonvie, Stefanie Sacks, Robin Farmanfarmaian, and Christy Whitman for taking the time to read my manuscript and providing such generous and thoughtful comments.

Thank you also to all of the parents and organizations, such as the Feingold Group and Sue Dengate, the founder of Food Additive

Network, for giving me a place to start looking for answers. And much gratitude for all of the scientists who work every day for hope and cures.

And finally, thank you to Dr. Russell Dahl, whom I stumbled upon like water in the desert and who said yes indeed you are heading in the right direction.

Thanks to my precious Knute for allowing me to tell your story. (And also, for your help in the editing process. I think you enjoyed using your eagle eye and red pen!) And the very most thanks go to my husband, Robert. For some reason you chose to fall in love with me after seven minutes and have been by my side ever since. When we hold hands, we have the whole world.

SECTION 1

What Is Calcium ATPase?

In this section, I will give you an introduction to Calcium ATPase and what to expect in the pages to follow. A roadmap, so to speak, as to how to approach this book. (Hint: You can focus on the sections that are applicable to you or your family.) I will then go into more detail about how Calcium ATPase works. This is the place to start before you peruse the chapters ahead.

CHAPTER ONE

WHAT'S THE BIG DEAL ABOUT CALCIUM ATPASE?

So what is the big deal about Calcium ATPase? Here's the short and snappy version: Calcium ATPase is the key mechanism that prevents calcium imbalance within our cells. When calcium levels remain too high for an extended period of time, problems occur. Calcium within our cells is separate from calcium in our teeth and bones—don't worry, I will explain all of this in the next chapter. But what I want to get across to you right away is that reduced Calcium ATPase levels are associated with many chronic diseases, including:

Obesity

- Calcium ATPase is responsible for approximately 40% of calories burned by your muscles and approximately 12–15% of total calories burned.
- Thyroid hormone stimulates Calcium ATPase, which is the reason it increases metabolism.
- Low levels of Calcium ATPase increase fat storage and reduce fat burning through its impact on fatty acid synthase and lipolysis. Studies show that there is an inverse relationship between Calcium ATPase and Body Mass Index (BMI). The fatter you are, the lower your Calcium ATPase levels.

Diabetes

- Calcium ATPase is critical to the production of insulin by the pancreas. When Calcium ATPase levels are low, pancreatic cells are damaged and less insulin is produced.
- High blood sugar associated with diabetes in turn reduces Calcium ATPase levels throughout the body. This reduction in Calcium ATPase is the primary driver of diabetes-related diseases, including heart disease, high blood pressure, eye damage, and nerve damage.

Cancer

- Calcium ATPase has an impact on the growth and differentiation of cancer cells, contributing to cancer's progression.
- Reduced Calcium ATPase is directly associated with breast, lung, colon, thyroid, skin, and blood cancers.

Blood Pressure

- Studies show that high blood pressure is associated with reduced Calcium ATPase levels. Reduced Calcium ATPase makes it harder for the arteries to relax. This results in constriction or narrowing of arteries, which leads to high blood pressure.

Heart Failure

- Heart failure occurs when the heart does not pump sufficiently and inadequate oxygen and nutrients are delivered to the cells. Calcium ATPase plays a key role in the ability of the heart to pump. Reduced Calcium ATPase is a hallmark of heart failure.

Atrial Fibrillation (AFib)

- Reduced Calcium ATPase in the heart is associated with AFib—which is the most common type of heart arrhythmia that occurs when the heart beats out of rhythm. This leads to increased risk of heart attack and stroke.

Blood Clots

- Calcium ATPase, through its stimulation by nitric oxide—a substance the body releases in response to platelet aggregation—is the primary mechanism through which platelet aggregation is reduced and the likelihood of blood clots goes down.

Brain Function

- Adequate Calcium ATPase is necessary in the brain to prevent excessive intracellular calcium levels. Excessive intracellular calcium levels can have a detrimental effect on many brain processes.
- Reduced Calcium ATPase levels and calcium dysregulation are known endpoints in Alzheimer's disease.
- Calcium ATPase plays an important role during neurodevelopment.

Inflammation

- When Calcium ATPase is reduced, mast cells release inflammatory substances. The substances released by mast cells are associated with numerous inflammatory conditions, including irritable bowel, arthritis, and other chronic conditions, including diabetes and cardiovascular disease.
- Reduced Calcium ATPase results in the magnification of a normal allergic response to a set trigger.

IF CALCIUM ATPASE IS SO IMPORTANT, WHY HAVEN'T I HEARD OF IT BEFORE?

Calcium ATPase was discovered in 1961, which seems like a long time ago, but in the scientific research arena, it's a relatively brief span of time. Over the last fifty-plus years, a tremendous body of research has been accumulating, including over 24,000 articles published in the scientific literature concerning Calcium ATPase. However, although researchers have been investigating Calcium ATPase and the enormous impact it has on our well-being, the results have not been widely shared, residing in individual silos of knowledge specific to their areas of expertise.

By its nature, scientific research is as much art as science. There is not one master planner directing the process from above. Rather, thousands of scientists are each focused on a tiny piece of the puzzle that is the human body in all areas of science, including Calcium ATPase. This is an essential part of the process of discovery—without intense focus in particular areas, there would be no data for grand conclusions. The downside, however, is that sometimes it takes longer to reach important connections.

Current research falls under three umbrellas:

1. Calcium ATPase and disease
2. Toxins and other substances that reduce Calcium ATPase
3. Molecules, nutrients, or gene therapy that increase Calcium ATPase levels

The goal of this book is to synthesize the research from all three areas so that you have the knowledge to make the best health choices possible for you and your family.

CHAPTER TWO

HOW CALCIUM ATPASE WORKS

When most people think of calcium, they think of dairy products, healthy bones, and osteoporosis, which is the gradual weakening of bones through the loss of calcium that tends to happen with age. While all of these popular associations with calcium are correct, they only scratch the surface. In fact, calcium is an integral component of the estimated one billion cellular chemical reactions that occur in all 37 trillion cells within your body . . . every single second.

Throughout this book, we will delve into the science of intracellular calcium, looking closely as if through a microscope. But first, to give you a better idea of what our subject matter is all about, let's get a bird's-eye view of intracellular calcium's function in day-to-day life.

Let's say you've finished the day's to-do list, settled into your favorite chair, and picked up this book to read. Without even looking, you reach up with your hand and arm to switch on the reading lamp. Turning your gaze and attention to the book cradled in your hand, you use your thumb and fingers to open to this page and begin to read. Now let's briefly describe this scene from the perspective of calcium's role and its regulation.

To make the mental decision that you have time to read this book, calcium levels rise at the terminal end of nerves in your brain, releasing neurotransmitters that trigger your excellent choice.

Sitting down utilizes calcium in the motor nerves to transmit the signal from the brain to the quadriceps muscles on the front of your thighs to contract and slowly lower you into the chair. Once you've settled in, the signal stops and the quads relax. At the same time the quads are working, other signals are sent to the abdominal muscles and the hip flexor muscles, causing them to contract so you can bend forward at the waist and into a sitting position. Simultaneously, the muscles in the back of your legs, your hips, and your back are relaxing to allow the movement to take place. All this happens while the biceps muscles in your upper arm keeps your elbow bent and flexor muscles in your hand keep the book held out in front of your chest. While still settling in, you decide a little more light would be better for reading. You reach up and over to the lamp by contracting the triceps muscles on the back of your arm along with some additional muscles in your back, while simultaneously relaxing the bicep muscle on the front of the arm and the pectoralis muscle on the front of your chest. You finish by closing your fingers around the knob and twisting the switch to the "on" position. All without glancing up.

As you open the book and start reading, tiny muscle contractions and relaxations are occurring continuously in the irises of both eyes to allow just the right amount of light to reach photoreceptor cells at the back of the eye. These cells utilize calcium and calcium regulators to send signals to the brain, which in turn utilizes even more calcium to process and learn the information in this book and think, "Wow, this is cool!"

Each of these physiological events is made possible by rising and lowering levels of calcium in your brain and nervous system, as well as rising and lowering levels of calcium in your muscles.

And there's so much more. For the entire time you've been reading, your heart has been beating, using tightly controlled increases and reductions in calcium levels to keep a steady pace of contractions and relaxations at about one beat or so per second. You notice the pot roast you put in the slow cooker earlier this morning is beginning to smell delicious because calcium has entered smell sensory cells. Your mouth even waters a bit when calcium triggers the production and release of digestive enzymes into your saliva.

A complete description of all that happens in our body in one minute, and how much of it depends on regulating the amount of calcium going into and out of cells, could fill a library—but that's not our goal. Instead, this is a quick glimpse at the marvelous inner workings of these bodies we call home in the context of intracellular calcium.

CALCIUM AND CALCIUM ATPASE

Next, let's consider what calcium actually is. Calcium is a chemical element, represented by the periodic table symbol *Ca*. It is an alkaline earth metal and the fifth most abundant element in the Earth's crust. As the ion Ca2+, it is also the fifth most abundant dissolved ion in seawater. Calcium is one of the basic ingredients with which the world is made, us included. In fact, *calcium is the most abundant mineral in the entire human body*, comprising about 1.5% of body weight, or 2.25 pounds in an average adult. A little over 99% of that total calcium weight resides in the bones and teeth. Our bodies combine calcium with other materials to make these structures both strong and resilient.

The other 1%, approximately one quarter of one ounce of calcium (about one-and-a half teaspoons), is not used in these hard structures. Referred to as *free calcium* by biologists, this calcium is not bound to anything and resides in the cells and bloodstream. The body uses this quarter ounce of free calcium to perform a wide variety of functions and chemical reactions. Though seemingly insignificant in quantity compared to the over two pounds found in our bones and teeth, it is still equivalent to 164,200,000,000,000,000,000,000,000 calcium ions (give or take a few). We are talking about a lot of really tiny stuff here.

Calcium ions play an essential role in a significant percentage of intracellular chemical reactions. The abilities to deliver calcium where it's needed and to remove it where it's not are critical functions, and our bodies have developed sophisticated ways to accomplish these tasks. The primary way is via an enzyme called Calcium ATPase (enzymes are proteins that act as catalysts to help initiate or speed up a chemical reaction, such as regulating the optimal levels of intracellular calcium).

Let's start with expanding on the calcium enzyme itself, Calcium ATPase.

- **Calcium**, we understand.
- **ATP** is another beast altogether: **Adenosine Triphosphate** (ATP) temporarily stores and transports chemical energy inside cells. It is produced by various cellular processes, most predominately in mitochondria. Think of it as little nuggets of energy.
- "**ase**" is a suffix denoting an enzyme. Most enzyme names end with it, and it indicates an action on the substrate in the name. Calcium ATPase, therefore, is an enzyme that moves calcium ions and requires energy (ATP) to perform the task.

INTRACELLULAR CALCIUM: THE NUTS AND BOLTS

Intracellular calcium is likely a new concept for you, so we'll start with the basics of what it is and how it functions.

Very simply, intracellular calcium is the calcium inside the cell's cytoplasm, which is the fluid that fills the inside of a cell. Every single cell in our body contains calcium, along with several other mineral ions and specialized structures such as organelles, mitochondria, and the nucleus (remember these terms from high school biology?). Without calcium and its proper function, the cell, and therefore you, could not exist.

While the presence of calcium is important, it's just as vital that you have the right amount of calcium inside your cells at any given moment. How much calcium each cell contains is finely calibrated according to need. The body maintains a very low amount of calcium inside the cell when it wants that cell to rest. When the body wants the cell to work, the nervous system signals the cell membrane to allow more calcium into the cytoplasm. To make the work happen quickly once the signal is received, our bodies maintain a very high concentration of calcium in the area outside the cells. This is called extracellular calcium. The ratio of extracellular to intracellular calcium ranges from 20,000:1 to 100,000:1. This immense pressure difference makes it easy to get lots of calcium into the cell when needed, like the pressure behind a water spigot when it's closed.

RISING CALCIUM LEVELS: A CALL TO ACTION

As we discussed above, when the body wants something to happen, it sends a signal to the cell to open the spigots that let calcium flow into the cytoplasm. This rise in intracellular calcium acts as a switch to initiate a particular function. Some examples of the actions intracellular calcium initiates are:

- In muscle cells, a rise in intracellular calcium triggers muscle contraction.
- In cardiac cells, a rise in intracellular calcium triggers the heart to beat.
- In neuronal cells, a rise in intracellular calcium triggers neurotransmitter release.
- In mast cells (a type of white blood cell), a rise in intracellular calcium triggers histamine release.

The spigots that allow calcium into the cell are transmembrane proteins called *calcium channels*. They open to allow calcium to flow into the cell when a signal, the stimulus, is received from the body. Some types of these calcium channels open by an electrical signal, others by a neurotransmitter molecule, and others when a hormone attaches. All of them close to stop the flow of calcium into the cells as soon as the signal stops.

Because these events and tasks need to be tightly controlled, our cells also have internal storage compartments to hold calcium. There is active coordination between how much calcium is released from these internal stores versus how much comes in from outside the cell. The technical term for these storage compartments is *vesicle*. The vesicles that store calcium are found in the *endoplasmic reticulum*, an organelle found in every cell in the body. (In addition to releasing calcium into the cytoplasm when signaled, the endoplasmic reticulum performs other calcium-triggered work such as insulin production.) In muscle cells, the additional storage compartments are found in the *sarcoplasmic reticulum*.

Here's a more nuanced layout of the process:

1. The body sends a signal to the cell to perform a task.
2. The signal causes calcium channels to open, and intracellular calcium levels rise.
3. The increased calcium causes the endoplasmic reticulum or sarcoplasmic reticulum to release its calcium stores.
4. Calcium levels rise to a certain level, and the cell completes the task.
5. The signals are removed, and calcium stops flowing into the cell.

LOWERING CALCIUM LEVELS: BACK TO BASELINE

Intracellular calcium is lowered back to baseline in four ways. The primary way is through sarco/endoplasmic reticulum Calcium ATPase (SERCA) and its transport of calcium ions into storage compartments within the cell. Two additional mechanisms to modulate intracellular calcium homeostasis are through the plasma membrane Calcium ATPase and the sodium-calcium exchanger, both of which transport calcium ions from the cytoplasm to the extracellular space outside of the cells. A fourth mechanism is the mitochondrial calcium transporter, whereby mitochondria take up calcium in small amounts to buffer the cell from excessive intracellular calcium levels. To further complicate matters, there is cross talk that occurs between all of these calcium regulation mechanisms! These latter three means are each worthy of careful attention and are the subject of considerable research as well, but the SERCA pathway will be my primary focus for this book.

Calcium ATPase is a sophisticated mechanism for returning elevated levels of intracellular calcium back to baseline levels, without which we would not be able to perform the relaxation phase of the contraction/relaxation cycle of any cell. In other words, if you shook hands with someone, you would never be able to release your grip without Calcium ATPase. Calcium ATPase performs one particular job: pumping the calcium released from the sarco/endoplasmic reticulum during the stimulus phase back to where it came from. In doing so, Calcium ATPase reduces intracellular calcium levels within the cell and ensures that adequate calcium is stored and available for the next stimulus. The level of calcium that is stored in the sarco/endoplasmic reticulum also plays a role in other important cell functions, such as insulin production, and cell growth and differentiation.

Inadequate levels of calcium in the sarco/endoplasmic reticulum cause *endoplasmic reticulum stress* (ER stress) and, similar to any other type of stress, reduces optimal performance. Over time, ER stress can cause a cascade of chemical reactions that impair cell function and can lead to cell death. This ER stress contributes to many diseases such as diabetes and Alzheimer's.

CHAPTER SUMMARY

1. Calcium is a key mineral that plays an important role in the body.
2. Most calcium (99%) is found in bone, teeth, and other hard structures.
3. A small amount of calcium is found in the cytoplasm of cells.
4. The calcium inside the cell is called intracellular calcium.
5. The levels of intracellular calcium control a wide range of events including muscle contraction, neurotransmitter release, and cardiac muscle function.
6. The level of intracellular calcium is primarily controlled by sarco/endoplasmic Calcium ATPase.
7. Inadequate levels of sarco/endoplasmic reticulum Calcium ATPase results in endoplasmic reticulum stress. Endoplasmic reticulum stress can lead to cell dysfunction and death, contributing to diseases such as diabetes and Alzheimer's.

SECTION 2

Diseases Associated with Reduced Calcium ATPase

Before we dive into Calcium ATPase and disease, I want to make it clear that reduced Calcium ATPase is neither the cause of every disease nor the only answer for optimal health. The body is wonderfully complex due to a multitude of factors, from genetics to mitochondrial health to inflammation, and the list goes on and on. For example, mitochondria produce the ATP that Calcium ATPase needs to function, but if there is not enough Calcium ATPase, the mitochondria compensate by taking up the excess calcium, which can lead to mitochondrial damage or death. Another example is inflammation. Reduced Calcium ATPase magnifies the body's inflammatory response, but at the same time inflammatory mediators reduce Calcium ATPase. What I can say unequivocally is that calcium levels within the cells are the on/off switch for every cell function. If these are out of balance, problems are likely to occur. It is in that spirit that I share the knowledge I have of this importance enzyme.

CHAPTER THREE

OBESITY

There are three major areas where weight control, specifically obesity, and Calcium ATPase intersect:

1. There is an association between obesity and reduced Calcium ATPase.
2. Calcium ATPase plays in important role in the metabolic rate.
3. The thyroid hormone's relation to Calcium ATPase levels.

CALCIUM ATPASE AND OBESITY

Over the last fifteen years, researchers have been connecting the dots between obesity and reduced Calcium ATPase levels. One key study published in 2003 examined the relationship between the body mass index (BMI) and Calcium ATPase levels. In general, a BMI of between 25 and 29 is considered overweight, whereas a BMI over 30 is considered obese. The study looked at thirty healthy women with BMIs between 20 and 40 who donated blood. A regression analysis was performed to determine the relationship between Calcium ATPase levels in their blood cells and BMI. There was a significant inverse correlation. In other words, the higher the BMI (the more overweight a person was), the lower the Calcium ATPase levels.[1]

1 Nasser J, Hashim S, LaChance P. Calcium and Magnesium ATPase activities in Women with varying BMIs. Obesity Research Volume 10 No 11 November 2004.

Animal studies have also shown a correlation between obesity and reduced Calcium ATPase levels. One such study divided sixty thirty-day-old rats into two groups. The control group was fed a normal diet. The second group, designated the obese group, was fed a high-calorie diet. As expected, the high-calorie diet resulted in excess weight gain as compared to rats in the control group, and by the fifteenth week, the obese rats weighed approximately 30% more than rats in the control group. Calcium ATPase levels were approximately 50% lower in the obese rats as compared to control rats.[2] A similar rat study investigated the effects of an eight-week exposure to an obesity-inducing diet. After just eight weeks, the rats fed this high-calorie diet were approximately 10% heavier. In concert, Calcium ATPase levels were approximately 40% lower in the obese rats as compared to rats in the control group.[3]

Although the relationship between Calcium ATPase and obesity is not fully understood, we know that obesity-inducing diets are associated with high blood sugar levels (hyperglycemia), and high blood sugar has been proven to reduce Calcium ATPase. (This is covered extensively in Chapter Four on Diabetes.)

Let's now take a look at the effect reduced Calcium ATPase has on our calorie-burning mechanism, the metabolism.

METABOLISM AND CALCIUM ATPASE

Our bodies use calories in myriad ways each day, and the metabolic rate is, in the most basic terms, the number of calories our body uses each day to function. Calcium ATPase uses energy (a.k.a. calories) in the form of ATP to fuel its pump. The food we eat goes through several metabolic pathways to form ATP and yield other important resources like amino acids and minerals. Calories are used when the ATP is utilized as energy within the cell. So, by definition, if you have inadequate levels of Calcium ATPase, the ATP energy (or calories burned) used by the body to regulate intracellular calcium will be lower than if you had normal levels of Calcium ATPase.

2 Lima-Leapoldo A, Leopoldo A, Silva D, do Nascimento A, Solome de Campos D, Luvizotto R, Oliveira S, Padovani C, Nogueira C, Cicogna A. Influence of Long Term Obesity on Myocardial Gene Expression. Doctoral Thesis submitted by Ana Paula Lima Leopoldo from Universidade Estadual Paulista 2012.

3 Huisamen B, Dietrich D, Bezuidenhout N, Lopes J, Flepisi B, Blackhurst D, Lochner A. Early cardiovascular changes occurring in diet-induced, obese insulin-resistant rats. Mol Cell Biochem (2012) 368:37–45.

Calcium ATPase burns calories in two distinct ways. One way is when Calcium ATPase pumps calcium into the sarcoplasmic or endoplasmic reticulum (SR or ER) to return intracellular calcium to resting levels. As we learned earlier, the sarcoplasmic reticulum is a calcium storage sac that serves almost the same function as the endoplasmic reticulum, but only to muscle cells. The second way calories are burned is when the Calcium ATPase pump functions without pumping calcium. This is called "uncoupled" transport and is quite normal. Fascinatingly enough, although the pumps are not transporting calcium, they are still using up ATP, which is released as heat.[4] You can imagine this process by thinking of a rental car shuttle van that transfers travelers from the airport to the rental car location. No matter what, the shuttle makes its round from terminal to terminal like clockwork. It may be that sometimes there are a lot of passengers to be transported, and other times the bus is empty. However, in both cases, the van still uses gas. This is analogous to uncoupled transport—energy is used despite no transport of calcium.

Calcium ATPase burns calories, whether pumping with calcium or pumping without calcium. Lower Calcium ATPase levels mean that with either method, fewer calories will be burned, resulting in a lower metabolic rate and making it more difficult to lose weight.

Let's touch on a bit of the research to back this up. Although Calcium ATPase is found in cells throughout the body, current research has focused on Calcium ATPase's metabolic impact in skeletal muscle and brown fat, which we explain below.

SKELETAL MUSCLE

Skeletal muscle simply refers to the muscles that enable us to move the various parts of our skeleton. Examples of these muscles include the biceps, triceps, quadriceps, and hamstrings. They account for between 40% to 50% of human body weight and are therefore quite relevant in terms of metabolism.

In one study, researchers looked at how much ATP was consumed by Calcium ATPase in rabbit skeletal muscle (rabbit skeletal muscle is a good approximation of human skeletal muscle). They

4 de Meis L, Arruda A, Carvalho DP. Role of Sarcoplasmic reticulum Calcium ATPase in thermogenesis. Biosci Rep 2005 Jun-Aug 25(3–4)181–90.

found that ATP consumption by Calcium ATPase in skeletal muscle was responsible for 40%–50% of calories burned by muscles at rest. Because the body is made up of more than just muscle, this translates into skeletal muscle Calcium ATPase being responsible for approximately 12%–15% of total body metabolism. Remember, this is just for the skeletal muscle—Calcium ATPase is found in every cell of the body, utilizing ATP, and hence calories.

The researchers concluded that Calcium ATPase could represent a potential target for interventions designed to treat obesity.[5]

BROWN FAT

Two types of fat exist in the body, brown and white. The key difference is that white fat is the net result of storing excess calories. In other words, when we eat more calories than we need, the excess is converted and stored as white fat. Brown fat's function, on the other hand, is not to store energy, but rather to generate heat by burning calories. Brown fat is an internal heater for the body. Calcium ATPase is the primary driver of heat production, because when Calcium ATPase pumps in an uncoupled manner (as we described above), the ATP energy is given off as heat rather than used to transport calcium. Brown fat is unique in that it has a greater proportion of uncoupled Calcium ATPase transport than other tissues, including skeletal muscle. A person who is overweight has less brown fat, proportionally, than a person of normal weight. Thus, there is less uncoupled Calcium ATPase activity, which results in a lower metabolic rate. Research evidences that adding more brown fat to mice increases their metabolism, which reduces white fat even when they are fed an obesity-inducing diet.[6]

---

5 Smith I, Bombardier E, Vigna C, Tupling A. ATP Consumption by sarcoplasmic reticulum Calcium ATPase accounts for 40–50% of resting metabolic rate in mouse fast and slow twitch skeletal muscle. PLOS 1 July 2013 Volume 8 Issue 7.

6 de Meis L., Brown adipose tissue CalciumATPase: uncoupled ATP hydrolysis and thermogenic activity. J Biol Chem 2003 Oct 24:278(43); 41856–61.

THYROID FUNCTION

The last piece of the puzzle in the relationship between obesity and Calcium ATPase is thyroid function. Most people are aware to some degree of the importance of thyroid hormone in metabolism. At a basic level, lower-than-normal thyroid hormone levels (hypothyroid) can lead to weight gain, whereas excessive thyroid hormone levels (hyperthyroid) can lead to weight loss. But what most people do not know is that the bridge between thyroid hormone and metabolism is Calcium ATPase.

As we discussed above, Calcium ATPase activity is accountable for a significant portion of the body's metabolism and, hence, calorie burn. It turns out that thyroid hormone increases Calcium ATPase levels, thereby increasing metabolic rate. Let's look at some research.

One study looked at skeletal muscle Calcium ATPase levels in eleven human subjects with hyperthyroid levels vs. a control group with normal thyroid levels. In hyperthyroid subjects, Calcium ATPase was approximately 40% higher than in the control group. So what did that mean for calorie burn? Hyperthyroid subjects had an energy expenditure rate approximately 40% greater than the control group. Researchers concluded that thyroid hormone, Calcium ATPase levels, and metabolism are significantly correlated.[7]

Another study compared the levels of skeletal muscle Calcium ATPase levels of hyperthyroid rabbits vs. the skeletal muscle Calcium ATPase levels of normal thyroid rabbits. Calcium ATPase levels in the hyperthyroid rabbits were four to five times greater than those obtained from control rabbits. In addition to increasing Calcium ATPase levels, hyperthyroidism increased the rate of uncoupled Calcium ATPase transport. In fact, depending on the muscle type, the rate of heat production by Calcium ATPase was four times greater in hyperthyroid skeletal muscle vs. normal thyroids.[8]

In hypothyroidism (as opposed to hyperthyroidism), as you would expect, there is a reduction in Calcium ATPase levels. One study looked at how a hypothyroid state would affect Calcium ATPase levels in the heart. It found that hypothyroid rats had 49%

7 Arruda AP, Da-Silva WS, Carvalho DP, De Meis L. Hyperthyroidism increases the uncoupled ATPase activity and heat production by the sarcoplasmic reticulum Calcium-ATPase. Biochem J. 2003 Nov 1; 375(Pt 3):753–60

8 de Meis L, Arruda AP, da-Silva WS, Reis M, Carvalho DP. The thermogenic function of the sarcoplasmic reticulum Calcium-ATPase of normal and hyperthyroid rabbit. Ann N Y Acad Sci. 2003 Apr; 986:481–8.

lower Calcium ATPase levels in the left ventricles of their hearts, which resulted in abnormal cardiac contraction and relaxation.[9] Another study confirmed these findings, reporting a 33% reduction in Calcium ATPase levels in hypothyroid rats as compared to a control group. This decline in cardiac Calcium ATPase resulted in symptoms of heart failure.[10]

With such a clear correlation between thyroid hormone levels and Calcium ATPase, it is important to do all you can to optimize the enzyme while working with your physician to deal with any hormonal imbalance when trying to lose weight.

Future Directions

- Recent research has shown that sarcolipin (a cellular protein) stimulates uncoupled Calcium ATPase activity, thereby increasing calories burned. Further research is needed, but the stimulation of sarcolipin may be a promising approach to treat obesity.[11]
- In 2020, researchers at Stanford and the University of California, San Diego reported that the use of a tiny implantable device that stimulates Calcium ATPase in fat cells increased metabolic rate by over 20%. In animal studies, mice with the implant device did not gain weight despite being fed an obesity-inducing diet.[12]

9 Montalvo D, Pérez-Treviño P, Madrazo-Aguirre K, González-Mondellini FA, Miranda-Roblero HO, Ramonfaur-Gracia D, Jacobo-Antonio M, Mayorga-Luna M, Gómez-Víquez NL, García N, Altamirano J. Underlying mechanism of the contractile dysfunction in atrophied ventricular myocytes from a murine model of hypothyroidism. Cell Calcium. 2018 Jun; 72:26–38

10 Tang YD, Kuzman JA, Said S, Anderson BE, Wang X, Gerdes AM. Low thyroid function leads to cardiac atrophy with chamber dilatation, impaired myocardial blood flow, loss of arterioles, and severe systolic dysfunction. Circulation. 2005 Nov 15; 112(20):3122–30

11 Maurya SK, Periasamy M. Sarcolipin is a novel regulator of muscle metabolism and obesity. Pharmacol Res. 2015 Dec;102:270–5. doi: 10.1016/j.phrs.2015.10.020. Epub 2015 Oct 30. PMID: 26521759; PMCID: PMC4684434.

12 Tajima K, Ikeda K, Tanabe Y, Thomson EA, Yoneshiro T, Oguri Y, Ferro MD, Poon ASY, Kajimura S. Wireless optogenetics protects against obesity via stimulation of non-canonical fat thermogenesis. Nat Commun. 2020 Apr 7;11(1):1730. doi: 10.1038/s41467-020-15589-y. PMID: 32265443; PMCID: PMC7138828.

CHAPTER SUMMARY

1. Obesity is associated with reduced Calcium ATPase levels.
2. This may be a result of hyperglycemia, which has a negative effect on Calcium ATPase.
3. Calcium ATPase is a key factor in metabolic rate or calories burned.
4. Calcium ATPase utilizes calories in two ways: it pumps "coupled" when it pumps calcium into the sarcoplasmic and endoplasmic reticulum, and it pumps "uncoupled" without the transport of calcium but while still utilizing energy.
5. The connection between the thyroid hormone and metabolic rate is Calcium ATPase. Higher thyroid hormone levels result in higher Calcium ATPase and higher metabolic rate, and lower thyroid hormone result in lower Calcium ATPase and lower metabolic rate.
6. Taking action to optimize Calcium ATPase levels is important to weight management.

CHAPTER FOUR

DIABETES

Calcium ATPase plays a role in diabetes in two critical ways:

1. Reduced Calcium ATPase has a negative effect on both insulin production and release.

2. High blood sugar associated with diabetes reduces Calcium ATPase, which results in a cascade of negative effects throughout the body.

At its most basic level, diabetes is the body's inability to regulate blood sugar levels, resulting in elevated blood sugar levels termed *hyperglycemia*. Hyperglycemia is the root of all diabetes-related health issues (we will explain why later in this chapter). Under normal circumstances, after you eat a meal, your blood sugar rises. The more carbohydrates you eat, the higher the blood sugar level. In response to the increase in sugar, also known as glucose, the pancreas secretes insulin. Insulin transports some of the glucose out of the blood and into cells and other parts of the body, bringing the blood glucose level back to a normal range. With diabetes, there is an inability of the pancreas to produce sufficient insulin, or the cells no longer respond to the insulin, which leads to hyperglycemia.

There are two types of diabetes, conveniently named Type 1 and Type 2. Type 1 diabetes (also referred to as *juvenile diabetes*) is when the pancreas is unable to produce sufficient insulin. Most Type 1 diabetes cases are diagnosed in children and young adults,[13] and as of yet, there is no known cause. Type 1 diabetes contains an autoimmune factor, in which the body attacks and destroys its own pancreatic beta cells, the cells responsible for producing insulin.[14] For the Type 1 diabetic patient, supplemental insulin is not an option, it's a requirement. The body simply does not produce enough, and without the additional insulin, the individual will die.

Type 2 diabetes (also referred to as *adult-onset diabetes*) is a different story. In this case, the pancreas initially produces sufficient insulin to control blood sugar levels, but various cells within the body, such as muscle, fat, and liver cells, stop responding to the insulin, developing what's called *insulin resistance*. The net result of insulin resistance is that sugar is not transported out of the blood into the cells, thus leading to high blood sugar (hyperglycemia). The pancreas responds to hyperglycemia by producing yet more insulin. However, over time, Type 2 diabetes sets in, wherein the insulin does not function effectively to maintain healthy blood glucose levels and the pancreatic cells become damaged. Type 2 diabetes is often associated with obesity and lack of exercise.

There is a multifaceted relationship between diabetes and Calcium ATPase. The information is dense and detailed, so we have divided it into three parts.

DIABETES PART ONE

INSULIN PRODUCTION AND RELEASE—THE CALCIUM ATPASE CONNECTION TO PANCREATIC BETA CELLS

The pancreas produces and releases the hormone insulin, the body's warrior to reduce excess blood sugar in our bodies. I call it a warrior because the negative effects of high blood sugar can be devastating (I'll delve into this in Part Three). The pancreas contains unique cells called *beta cells* whose function is to produce, store, and release insulin. When blood sugar rises, beta cells respond by secreting insulin from a readily available pool.

13 https://www.diabetes.org/.

14 Cnop M, Welsh N, Jonas J, Jorns A, Lenzen S, Eizirik. Mechanisms of Pancreatic Beta-Cell Death in Type 1 and Type 2 diabetes. Diabetes, Vol. 54, Supplement 2, December 2005.

Simultaneously, the release of insulin triggers insulin production, which ensures there will be an adequate supply of insulin for future needs.[15]

As in all other cells, the rise of intracellular calcium within a beta cell acts as a signal that triggers a response. In beta cells, the rise in intracellular calcium triggers insulin storage vesicles (similar in function to the endoplasmic reticulum storage sacs) to release insulin into the bloodstream. After insulin is released, Calcium ATPase returns intracellular calcium to resting levels by pumping calcium into the endoplasmic reticulum, where it will be available for release after the next stimulus.

Thus, Calcium ATPase serves two functions: It reduces intracellular calcium to resting levels, which is critical because prolonged elevated intracellular calcium results in cell damage or death, and it refills the endoplasmic reticulum with calcium. This is important regarding insulin because it ensures there will be adequate calcium stored for future insulin need.

The amount of calcium stored in the endoplasmic reticulum doesn't just impact insulin release, it also affects the multistep process of the manufacture of insulin itself. The calcium stored in the endoplasmic reticulum must remain at a certain level or it triggers *endoplasmic reticulum stress*, upon which a number of negative events occur within the cell. Just as having a panic attack prevents you from performing tasks as well as usual, when endoplasmic reticulum stress is present, the beta cells' production of insulin is compromised. The stages of insulin formation get out of whack and a much larger percentage of immature insulin material accumulates, which disturbs the natural development and supply of insulin ready to be released. In other words, the assembly line for the manufacture of insulin is disrupted. On top of that, the insulin that is produced has a greater chance of being defective. Furthermore, under conditions of endoplasmic reticulum stress, hormones are released that can lead to beta cell damage and death.

Let's look at a study that demonstrates the effects of low Calcium ATPase on blood sugar levels and insulin production.

15 Lipson K, Fonseca S, Ishigaki S, Nguyen L, Fosse, Bortell R, Rossini A, Urano F. Regulation of insulin biosynthesis in pancreatic beta cells by an endoplasmic reticulum-resident protein kinase IRE1. Cell Metabolism 4, 245–254, September 2006.

Researchers genetically manipulated one group of mice to have reduced Calcium ATPase levels, approximately 50% lower than the control group. Both groups ate a high-calorie diet for sixteen weeks. Predictably, both groups gained weight. However, after the weight gain occurred, significant differences were found between the two groups' ability to maintain normal blood sugar levels. In response to a glucose challenge (typically a five-hour fast followed by a glucose injection designed to measure glucose tolerance, or the body's ability to respond to elevated blood sugar levels), serum insulin levels were 55% lower in the reduced Calcium ATPase mice as compared to controls. Due to the reduced insulin levels, it took significantly longer for the reduced Calcium ATPase mice to restore blood sugar to normal. There was also a significant difference in the amount of calcium stored in the endoplasmic reticulum. (Remember, reduced Calcium ATPase levels by definition means less calcium is being pumped into the endoplasmic reticulum.)

As we mentioned above, endoplasmic reticulum calcium levels have an impact on insulin production, processing, and packaging, so it was no surprise that the total amount of insulin stored in the beta cells of the reduced Calcium ATPase mice was 20% lower. Furthermore, the insulin maturation process was abnormal, with the percentage of immature granules two times higher in the reduced Calcium ATPase mice. There was also a significantly higher amount of defective insulin crystallization and packaging (the important final steps in making mature insulin and getting it ready to release as needed). To top it off, reduced Calcium ATPase mice had reduced beta-cell growth and increased beta-cell death. The study concluded that strategies aimed at improving Calcium ATPase function could be viable strategies to improve glucose stability.[16]

16 Tong X, Kono T, Anderson-Baucum E, Yamamoto W, Gilon P, Lebeche D, Day R, Shull G, Evans-Molina C. SERCA2 Deficiency Impairs Pancreatic Beta-Cell Function in Response to Diet-Induced Obesity. Diabetes 2016; 65:3039–3052.

DIABETES PART TWO

HIGH BLOOD SUGAR—BAD NEWS FOR CALCIUM ATPASE

Prolonged hyperglycemia, or high blood sugar, that occurs with diabetes also inhibits Calcium ATPase in numerous body organs. This is of critical importance, as the reduction in Calcium ATPase contributes to many diseases associated with diabetes, including heart disease, hypertension, sensory neuropathy, and cataracts. There are two primary ways that high blood sugar negatively affects Calcium ATPase: the formation of advanced glycation end products that disable Calcium ATPase and free radicals that damage Calcium ATPase.

Advanced Glycation End Products (AGEs)

Advanced glycation end products (AGEs) are created from the reaction between sugar and proteins. So, when blood sugar levels are elevated, such as in diabetes, there is greater likelihood that AGEs will be formed. Like mistletoe in a tree, AGEs are parasites to the proteins they attach themselves to, ultimately disabling their function. For our purposes, what's important is that Calcium ATPase is a protein, and having diabetes increases the formation of AGEs on Calcium ATPase, thereby reducing its ability to transport calcium.

Let's look at some research.

In one study, researchers used a test called mass spectrometry to examine Calcium ATPase proteins in diabetic rats. The test revealed significant numbers of AGEs on the Calcium ATPase in the diabetic rats as compared to nondiabetic rats. Not surprisingly, the number of AGEs attached to Calcium ATPase was reduced when the diabetic rats were treated with insulin for two weeks. The insulin lowered blood sugar levels and therefore AGEs.[17]

Another study looked at the impact AGE inhibitors would have on Calcium ATPase levels in diabetic rats. The researchers measured the Calcium ATPase levels of three groups of rats: diabetic, diabetic and treated with an AGE inhibitor (which would inhibit the formation of AGE), and a control group. Untreated diabetic rats had Calcium ATPase levels 64% lower than rats in the control group. The AGE inhibitor-treated rats also had lower Calcium ATPase levels

17 Bidasee K, Zhang Y, Shao C, Wang M, Patel K, Dincer U, Besh H. Diabetes increases formation of advanced glycation end products on Sarco(endo)plasmic Retciulum CalciumATPase. Diabetes, Vol 53, Feb 2004.

than the rats in the control group; however, their Calcium ATPase levels were still 66% higher than those of nontreated diabetic rats. This study demonstrated the possibility of improving Calcium ATPase levels in diabetic rats through the inhibition of AGEs. It's not a stretch to assume the same is true of humans.

Doctors use A1-C levels to monitor blood sugar over time. This test measures the amount of glycated hemoglobin protein in the blood. The hemoglobin protein is glycated in a similar fashion as Calcium ATPase, the sugar in the blood attaching itself to the protein, in this case the hemoglobin. A recent study found that A1-C levels and Calcium ATPase are inversely correlated. The higher the A1-C, the lower the Calcium ATPase levels. High blood sugar is the underlying connection.[18]

High Blood Sugar and Oxidation

The second way hyperglycemia reduces Calcium ATPase is through the creation of free radicals. Hyperglycemia has been shown to generate reactive oxygen species (ROS). ROS are molecules that have an unpaired electron in one of their orbits, called a free radical. This radical state, with the unpaired electron wanting to be paired, makes the molecule both unstable and very reactive. It wants to donate this electron, thereby becoming an oxidant and oxidizing the other molecule, essentially causing it to rust. A buildup of ROS in cells may cause damage to DNA, RNA, and proteins and cause cell death. Calcium ATPase is a protein that can be damaged by ROS created in high blood sugar states.[19]

18 Galindo-Hernandez O, Machado-Contreras JR, Martinez-Corella R, Romero-Garcia T, Vazquez-Jimenez JG. Inverse correlation between levels of glycated haemoglobin and expression levels of SERCA protein in Mexican patients with type 2 diabetes mellitus. Arch Med Sci. 2020 Aug 10;16(5):1226–1228. doi: 10.5114/aoms.2020.97970. PMID: 32864012; PMCID: PMC7444701.

19 https://www.ncbi.nlm.nih.gov/pmc/articles/PMC2944313/.

DIABETES PART THREE

DIABETES-RELATED DISEASES AND CALCIUM ATPASE

As we learned in the previous section, high blood sugar reduces Calcium ATPase through glycation and oxidation. This reduction in Calcium ATPase causes damage throughout the body and is responsible for many diseases associated with diabetes. We will look more closely at several of these disease states.

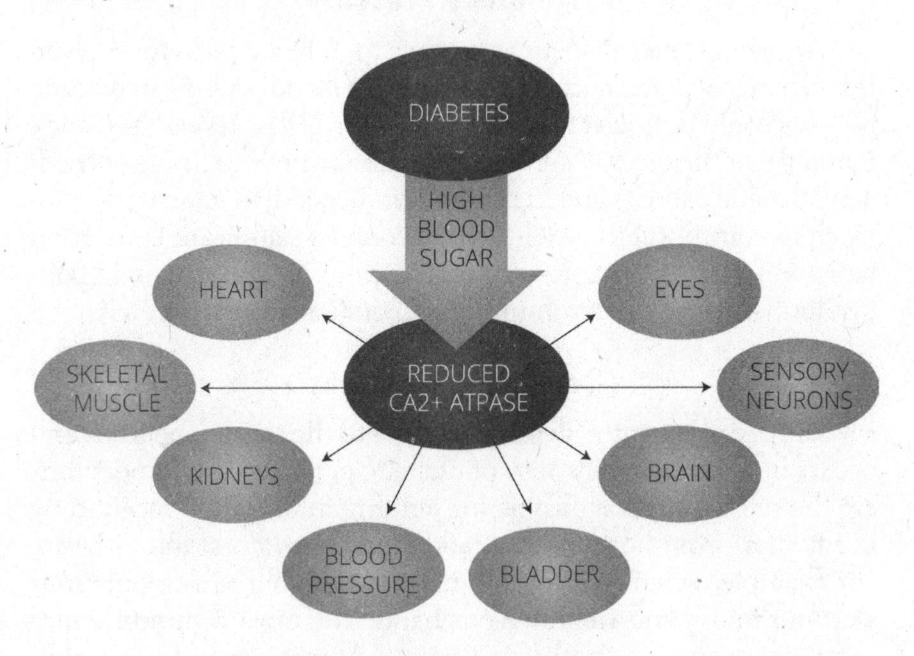

Heart Disease

Heart disease is the leading cause of death in diabetic patients. In fact, two out of every three people with diabetes die of heart disease. The specific term used to describe this cardiac dysfunction is *diabetic cardiomyopathy*. Although given a distinct name, diabetic cardiomyopathy is essentially heart failure, which I will cover in more detail in Chapter Ten.

In terms of research, numerous studies have reported reduced Calcium ATPase levels in the hearts of diabetic subjects. For example, in a direct comparison of healthy rat cardiac muscle versus that of diabetic rat cardiac muscles, one study found that six weeks after they induced diabetes, the diabetic rats had cardiac Calcium ATPase levels approximately 50% lower than the nondiabetic rats. The reduced levels of Calcium ATPase in the diabetic rats correlated with reduced contraction and relaxation of the heart muscle. Thus, reduced cardiac Calcium ATPase levels in diabetic patients are a causal factor in diabetic cardiac dysfunction.[20]

High Blood Pressure

Two out of three diabetics also have high blood pressure.[21] Given this fact, researchers wanted to learn if high blood pressure in diabetic patients could be linked to reduced Calcium ATPase levels. One study found that Calcium ATPase levels were lower in both insulin-dependent diabetic patients and in non-insulin-dependent patients as compared to control subjects. Researchers found a significant correlation between systolic blood pressure and Calcium ATPase activity in both insulin-dependent and non-insulin-dependent patients.[22]

Sensory Neuropathy

Sensory neuropathy denotes damage to the sensory neurons and occurs in approximately 50% of diabetic patients. Sensory neurons are the nerve cells responsible for sending information received by our bodies (from both internal and external sources) to our brain. For example, when we touch something hot, our sensory neurons alert our brain, and we move our hand. The most common symptoms of sensory neuropathy are tingling, numbness, and a sensation of "pins and needles" in hands, feet, fingers, or toes. In addition to being uncomfortable and sometimes painful, sensory neuropathy can result in a reduced ability to sense pain or extreme temperatures. Because of this reduced sensory awareness, patients may not be aware of blisters, cuts, or burns, which can become infected or be more serious than imagined. This is further compounded by the fact that diabetes alone causes reduced blood flow to the feet, which means slower healing times. Infections can spread to the bones,

---

20 Netticaden T, Temsah R, Kent A, Elimban V, Dhalla N. Depressed levels of Calcium Cycling Proteins May Underlie Sarcoplasmic Reticulum Dysfunction in Diabetic Heart. Diabetes, Vol 50, September 2001.

21 American Diabetic Association Website.

22 Spieker C, Fischer S, Zierden E, Schluter H, Tepel M, Zidek W. Cellular Calcium ATPase Activity in Diabetes Mellitus. Horm metab Res. 26(1994) 544–547.

which by then may be impossible to treat. Tragically, this condition can result in lower leg amputation, and in North America alone, 90,000 diabetes-related amputations are performed each year.[23]

In neurons, changes in intracellular calcium are responsible for neurotransmitter release and the transmission of the nerve signal. Thus, calcium regulation is central to the neuron's function. In diabetic sensory neurons, the calcium regulatory system is impaired, which is reflected in prolonged significant elevation of intracellular calcium levels. As in other cells, elevated intracellular calcium damages the cell and in this case causes the disintegration of the neuron's axon, which is the part of the nerve cell that carries information to other parts of the body.[24] In addition, elevated calcium reduces the neuron's ability to regrow and regenerate.[25]

To delineate what role Calcium ATPase plays in the calcium dysregulation found in sensory neuropathy, researchers looked at sensory neurons from normal and diabetic rats. They found that Calcium ATPase activity was reduced by more than twofold in the diabetic rats. To further confirm that reduced Calcium ATPase was central to sensory neuropathy, the researchers took sensory neurons from normal rats and treated them with a Calcium ATPase inhibitor, thereby blocking Calcium ATPase from working. When normal rats were treated with the Calcium ATPase inhibitor, they exhibited the same symptoms of diabetic neuropathy calcium dysregulation as the diabetic rats.[26]

Vision Problems

Cataracts are common among diabetic patients, with 66% of diabetics afflicted. While surgery is an option as a cure, diabetics have a greater risk of complications (including blindness) due to such factors as difficulties in dilating pupils and the development of diabetic macular edema. Therefore, substantial research has been

23 Verkhratsk A, Fernyhough P. Mitochondrial malfunction and Calcium dyshomeostasis drive neuronal pathology in diabetes. Cell Calcium (2008) 44; 112–122.
24 Coleman M., Axon Degeneration Mechanisms: Commonality amid Diversity. Nature Reviews: Neuroscience: Volume 6 Nov 20015 889–897.
25 Brandt PC, Sisken JE, Neve RL, Vanaman TC. Blockade of plasma membrane calcium pumping ATPase isoform I impairs nerve growth factor-induced neurite extension in pheochromocytoma cells. Proc Natl Acad Sci U S A. 1996 Nov 26;93(24):13843-8. doi: 10.1073/pnas.93.24.13843. PMID: 8943023; PMCID: PMC19443.
26 Zherebitskaya E, Schapansky J, Smith D, Ploeg R, Solovyova N, Verkhratsky A, Fernyhough G. Sensory neurons derived from diabetic rats have diminished internal Calcium stores linked to impaired re-uptake by the endoplasmic reticulum. Asn Neuro 4(1); e00072. Doi:10.1042/AN201110038.

conducted to determine which factors contribute to the formation of cataracts in diabetic patients with the goal of developing methods to reduce the likelihood of cataract formation.

Researchers have identified several different pathways contributing to cataract formation, including a sustained elevation of intracellular calcium levels in both animal and human models. Intracellular calcium is significantly elevated in only the opaque areas where cataracts are formed, whereas the surrounding clear areas have near normal intracellular calcium levels. The increased intracellular calcium leads to the activation of a process that ends in the loss of lens proteins. The loss of lens proteins plays a pivotal role in cataract formation.[27] Calcium homeostasis is essential to cataract prevention, and Calcium ATPase is a primary mechanism for doing this.

One study compared Calcium ATPase levels in diabetic rat lenses vs. a control group and found that Calcium ATPase levels were more than 80% lower in diabetic rats vs. the control group. These differences resulted in increased intracellular calcium levels, which were associated with cataract formation in the diabetic animals.[28] We can clearly see (pun intended) that decreased Calcium ATPase levels can adversely affect our vision through the formation of cataracts.

And More . . .

In addition to the areas mentioned above, the brain, kidneys, skeletal muscle, and bladder have also been shown to have reduced Calcium ATPase in diabetic patients. These reduced Calcium ATPase levels result in various organ-specific problems.

27 Duncan G. Calcium cell signaling and cataract: role of the endoplasmic reticulum. Eye (1999) 13, 480–483.
28 Sai Varsha M, Raman T, Manikandan R. Inhibition of diabetic-cataract by vitamin K1 involves modulation of hyperglycemia-induced alterations to lens calcium homeostasis. Experimental Eye Research 128 (2014) 73–82.

FUTURE DIRECTIONS

There are some exciting research findings regarding targeting Calcium ATPase in the treatment of diabetes. Here are just a few:

- Researchers at the Icahn School of Medicine at Mt. Sinai reported that diabetic mice treated with a Calcium ATPase activator had markedly lower fasting blood glucose and improved glucose tolerance, showed a significant reduction in adipose tissue, and had an increased metabolic rate as compared to diabetic mice with no treatment. Amazingly, the mice treated with this molecule maintained normal blood sugar levels more than six weeks after stopping the treatment. Parallel to the benefits in blood sugar control, the treated mice also had significant improvement of cardiac performance and mitochondrial function.[29]
- Researchers at Harvard reported that increasing the level of Calcium ATPase in the livers of obese and diabetic mice through a gene transplant alleviated endoplasmic reticulum stress, increased glucose tolerance, significantly reduced blood glucose levels, and reduced triglycerides. Importantly, in lean nondiabetic mice, the gene transplant did not lead to hypoglycemia, which indicated that activation of Calcium ATPase is a safe treatment target.[30]
- In another study, researchers reported that increasing the amount of Calcium ATPase in skeletal muscle through gene transplant in diabetic mice had a significant therapeutic effect on diabetic myopathy (muscle injury due to inflammation that progresses into degenerative muscle damage). In fact, the gene transfer was more effective than stem cell treatment.[31]

---

29 Kang S, Dahl R, Hsieh W, Shin A, Zsebo KM, Buettner C, Hajjar RJ, Lebeche D. Small Molecular Allosteric Activator of the Sarco/Endoplasmic Reticulum Ca2+-ATPase (SERCA) Attenuates Diabetes and Metabolic Disorders. J Biol Chem. 2016 Mar 4;291(10):5185-98. doi: 10.1074/jbc.M115.705012. Epub 2015 Dec 23. PMID: 26702054; PMCID: PMC4777852.

30 Park SW, Zhou Y, Lee J, Lee J, Ozcan U. Sarco(endo)plasmic reticulum Ca2+-ATPase 2b is a major regulator of endoplasmic reticulum stress and glucose homeostasis in obesity. Proc Natl Acad Sci U S A. 2010 Nov 9;107(45):19320–5. doi: 10.1073/pnas.1012044107. Epub 2010 Oct 25. PMID: 20974941; PMCID: PMC2984194.

31 Zickri MB, Sadek EM, Fares AE, Heteba NG, Reda AM. Effect of Stem Cells, Ascorbic Acid and SERCA1a Gene Transfected Stem Cells in Experimentally Induced Type I Diabetic Myopathy. Int J Stem Cells. 2020 Mar 30;13(1):163–175. doi: 10.15283/ijsc18066. PMID: 32114738; PMCID: PMC7119208.

- In a 2017 study, researchers reported that jaceosidin (a natural compound from the Japanese mugwort plant) triggered Calcium ATPase upregulation, enhanced insulin sensitivity, and decreased endoplasmic reticulum stress in diabetic mice. Furthermore, jaceosidin significantly suppressed blood glucose levels, improved glucose tolerance, and lowered body weight.[32]

---

32 Ouyang Z, Li W, Meng Q, Zhang Q, Wang X, Elgehama A, Wu X, Shen Y, Sun Y, Wu X, Xu Q. A natural compound jaceosidin ameliorates endoplasmic reticulum stress and insulin resistance via upregulation of SERCA2b. Biomed Pharmacother. 2017 May;89:1286–1296.

CHAPTER SUMMARY

1. Increased intracellular calcium triggers the beta cells to release insulin.
2. After insulin release, intracellular calcium levels are returned to normal levels by Calcium ATPase.
3. Calcium ATPase also ensures that there is an adequate amount of calcium in the endoplasmic reticulum that is necessary for insulin production.
4. If endoplasmic reticulum calcium stores are inadequate, endoplasmic reticulum stress occurs and insulin production is reduced, which may lead to cell death.
5. High blood sugar has a negative effect on Calcium ATPase.
6. When blood sugar levels are high, advanced glycation end products (AGE) are formed.
7. AGE occurs when a sugar molecule attaches to a cellular protein, rendering it useless.
8. Sugar binds to Calcium ATPase, resulting in AGE. This prevents Calcium ATPase from doing its job.
9. In addition, high blood sugar results in the production of free radicals, which can result in damage to Calcium ATPase.
10. Reduced Calcium ATPase levels contribute to many diabetic-related complications including heart disease, high blood pressure, sensory neuropathy, and cataracts.

CHAPTER FIVE

CANCER

In this chapter, I cover the relationship between Calcium ATPase and several distinct, common, and potentially deadly varieties of cancer.

Numerous studies have documented a relationship between calcium dysregulation and cancer. In particular, research has demonstrated that calcium dysregulation is involved in a key characteristic of cancer: abnormal cell differentiation (which I will explain below). Because Calcium ATPase plays a major role in calcium regulation, much of this research has examined the relationship between Calcium ATPase and cancer.

Let's begin with an overview of the cancer/Calcium ATPase connection.

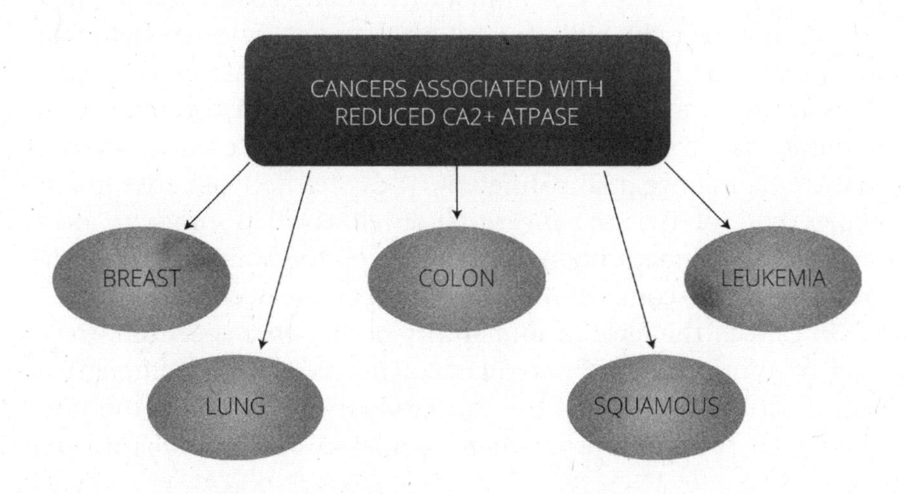

FUNDAMENTAL CONCEPTS WITHIN CALCIUM ATPASE AND CANCER

Cancer, at its very root, can be defined as the existence of abnormal cells that multiply excessively and invade nearby tissue. What are abnormal cells? Cancer cells do not look the same under a microscope as normal cells, because they are not "differentiated" as fully as normal cells. They are to varying degrees "undifferentiated." Still vague? Let's look at this in detail.

Cell differentiation describes the process by which immature cells become mature cells with specific functions. As they mature, normal cells *differentiate* into specialized cells. Each of our various types of tissue maintains a reservoir of *pluripotent cells*, also known as *stem cells*, which are cells that can become any other type of cell in the body. These stem cells divide to form new cells as replacements are needed. When a stem cell divides into two cells, both those cells are termed *daughter* cells. Depending on need, one daughter cell will begin the process of differentiating into a specialized mature cell and will have a limited ability to replicate thereafter. That latter point is important. The other daughter cell will remain a stem cell in order to be able to repeat the process as required. In times of greater need, genetic signaling can trigger both daughters to specialize.

It is during this genetic signaling process that damaged DNA can prevent a cell from completely maturing. Cancer cells are those cells that do not fully differentiate into mature specialized cells; they remain immature. Cell immaturity is dangerous, because those cells do not have the built-in limit on their ability to replicate that mature cells do. Therefore, these cells can proliferate more rapidly than normal and, unfortunately, in a manner that confounds our immune system's ability to recognize and eliminate them. As seen in the image above, many different types of cancer can arise in our bodies, and the type of cancer that develops when things go awry is directly dependent upon the type of cell the body was trying to create when the compromised genetic signaling occurred.

In cancer, the level of immaturity occurs on a spectrum. Cells may be more or less differentiated. The level of cell differentiation describes how much the cancerous tissue looks like the normal tissue surrounding it when compared under a microscope.

Differentiation is used in tumor grading systems, with tests and specific measurements differing between particular cancers. However, in general, the important marker for cancer staging is the amount of cell differentiation.

Highly differentiated cancer cells look a lot like normal cells and are usually slow-growing due to a greater limit on their ability to divide and reproduce. *Moderately differentiated* cancer cells look less like normal cells and, correspondingly, grow more rapidly. *Poorly differentiated* cancer cells do not look at all like normal cells and are fast-growing or "aggressive." One way to look at the immaturity or differentiation spectrum of cancer cells is that the further along in the maturation process (the more fully differentiated and normal-looking the cancer cell is), the more limited its ability to replicate. A poorly differentiated cancer cell is one that did not get very far in the process of maturing, which stunts its replication limits and allows it to grow more aggressively.

Calcium ATPase impacts cell differentiation because the level of calcium stored in the endoplasmic reticulum plays a crucial role in the all-important cell differentiation. Research shows that differentiated cells have greater Calcium ATPase levels than moderately differentiated cells, and in poorly differentiated cancer cells, Calcium ATPase levels can be completely absent. In other words, the level of Calcium ATPase is inversely correlated with cancer staging. The lower the levels of Calcium ATPase, the higher the stage of cancer, and thus the deadlier the cancer.

CALCIUM ATPASE AND DIFFERENT TYPES OF CANCER

Development of cancer is now thought of as a multistep process in which cells are driven sequentially through a series of biological mutations that culminate in cancer formation. Reduction in Calcium ATPase is one of these factors. Below, I detail research concerning Calcium ATPase in lung, squamous cell, colon, and breast cancer.

LUNG CANCER

For a start, researchers compared Calcium ATPase levels and endoplasmic reticulum calcium levels in cancerous lung tissue versus normal lung tissue. They found that Calcium ATPase levels were significantly reduced in cancer cells vs. normal cells. In small-cell lung cancer, Calcium ATPase levels were 55% lower than in normal cells, and in non-small-cell lung cancer, there was a 30% reduction in Calcium ATPase levels as compared to normal cells.

Researchers also found a direct correlation between reduced Calcium ATPase levels and endoplasmic reticulum calcium levels. Endoplasmic reticulum calcium stores were 60% lower in small-cell lung cancer and 30% lower in non-small-cell lung cancer as compared to normal cells. When calcium levels in the endoplasmic reticulum become too low and stay that way for too long, it causes what is termed "endoplasmic reticulum stress." In cancer, endoplasmic reticulum stress interferes with normal cell growth and differentiation.

Researchers then went one step further to see what effect a Calcium ATPase inhibitor would have on the proliferation of undifferentiated cells. After the application of a Calcium ATPase inhibitor to lung cancer cells for twenty-four hours—effectively shutting off the Calcium ATPase pump and thereby reducing endoplasmic reticulum calcium stores—cancerous cells proliferated by approximately 30% in non-small-cell cancer cells and approximately 20% in small-cell cancer cells. Remember, accelerated proliferation of undifferentiated cells is responsible for tumor growth. The researchers suggested that the differences in calcium regulation between cancerous lung cells and normal lung cells might be a new target for diagnostic or therapeutic approaches.[33]

A second set of researchers looked at it from a different angle. In human lung cancer patients, was there a genetic mutation in the ATP2A2 gene, the gene responsible for the encoding of Calcium ATPase? Could inefficient Calcium ATPase production and regulation be leading to cancer?

They analyzed tumor DNA of ninety-five patients with lung cancer and compared it to the DNA of normal (noncancerous) patients,

33 Bergner A, Kellner J, Tufman A, Huber R. Endoplasmic reticulum Ca2+-homeostasis is altered in small and non-small cell lung cancer cell lines. Journal of Experimental and Clinical Cancer Research 2009, 28:25.

looking for genetic mutations and the levels of ATP2A2 gene in each group. Lung cancer patients showed a significant increase of genetic mutations (in other words, abnormal genetic sequences) in the ATP2A2 gene. These mutations correlated with reduced expression of the ATP2A2 gene. The conclusion of the study was that reduced Calcium ATPase might play a role in increased susceptibility to the development of lung cancer.[34]

SQUAMOUS CELL CANCER

Squamous cell cancer is, as the name implies, a cancer found in a particular type of cell called a squamous cell (*squama* is Latin for fish scale, and squamous cells have a flat, fish-scale shape).

Overall, squamous cell cancers are common. In fact, almost half of all Americans age sixty-five and older will develop either basal or squamous skin cancer.[35]

Researchers studied that same ATP2A2 gene that encodes for Calcium ATPase. Specifically, they genetically modified the DNA in fourteen mice to decrease levels of the ATP2A2 gene. As a result, the genetically altered mice's Calcium ATPase levels were reduced 66% in the skin, 80% in the tongue, and 82% in the stomach.

Researchers then determined the susceptibility of these genetically modified mice to developing squamous cell tumors as they aged (in mouse terms, from fifty to eighty weeks). Thirteen out of the fourteen mice developed squamous cell tumors. Many of those thirteen mice had numerous tumors. This included squamous cell tumors in the oral cavity, tongue, esophagus, stomach, and skin. In all, thirty squamous cell tumors were identified among the fourteen genetically modified mice.

Age-matched normal mice did not develop a single tumor. In fact, squamous cell tumors are extremely rare among mice. According to the national toxicology program, there were only twenty-five cases of squamous cell tumors among 4,900 mice as compared to thirteen among fourteen in the genetically modified mice! The researchers concluded that Calcium ATPase deficiency

34 Korosec B, Glavac D, Rott T, Ravnik-Glavac M. Alterations in the ATP2A2 gene in correlation with colon and lung cancer. Cancer Genetics and Cytogenetics 171(2006) 105–111.
35 https://www.skincancer.org/skin-cancer-information/skin-cancer-facts.

predisposes mice to squamous cancer and may be a primary initiating event in cancer.[36]

Another group of researchers looked at the ATP2A2 gene in human patients with oral squamous skin cancer. The study involved tissue samples from three groups: fifty-two patients with oral squamous carcinomas, thirty-two patients with oral precancerous skin lesions, and eight patients with cancer derived from available oral squamous carcinoma cell lines. Normal tissue samples were utilized as the control group. In comparison to normal tissue, researchers found a 42% reduction of the ATP2A2 gene in oral squamous carcinomas and a 34% reduction of precancerous lesions. The reduced ATP2A2 gene encodes for Calcium ATPase, so it makes sense that a reduction in ATP2A2 would result in reduced Calcium ATPase. The researchers concluded that the reduced regulation of ATP2A2 may play a role in oral cancer development and can be seen as a "precipitous" event in oral cancer.[37]

COLON CANCER

Colon cancer and its broader sibling, colorectal cancer (including either the colon or rectum), is the second-most deadly form of cancer in America. On average, about one-third of those diagnosed will not survive past the five-year mark.[38]

Researchers measured Calcium ATPase levels in colon samples from twenty-seven patients with carcinomas, eight patients with adenomas (precancerous lesions that can turn into carcinomas), and a control group with normal colon tissue. In normal tissue, Calcium ATPase increased as the cells differentiated, which is what you would expect, as adequate endoplasmic reticulum stores are essential in normal cell differentiation. In the precancerous adenomas, Calcium ATPase levels were reduced compared to normal colon tissue. And in colon carcinomas, Calcium ATPase was nearly absent in poorly differentiated tumors (associated with the most malignant stage of cancer). The researchers concluded

36 Liu L, Bovivin G, Prasad V, Periasamy M, Shull G. Squamous cell tumors in mice heterozygous for a null alle of ATP2A2, encoding the sarcoplasmic reticulum Calcium ATPase isoform 2 Ca2+pump. The Journal of Biological Chemistry, Vol 276, No.29, Issue of July 20 pp 26737-26740 2001.

37 Endo Y, Uzawa K, Mochida Y, Shiba M, Bukawa H, Yokoe H, Tanzawa H. Sarcoplasmic reticulum Calcium ATPase Type 2 downregulated in human oral squamous cell carcinoma. Int J. Cancer; 110,225-231 (2004).

38 https://seer.cancer.gov/statfacts/html/colorect.html.

that intracellular calcium homeostasis becomes increasingly abnormal during colon cancer as reflected by deficient Calcium ATPase expression.[39]

Another study looked at the levels of ATP2A2 gene in patients with colon cancer and compared the results to the DNA of normal (noncancerous) patients. Colon cancer patients showed a significant increase of genetic mutations (in other words, abnormal genetic sequences) in the ATP2A2 gene. These mutations were correlated with reduced expression of the ATP2A2 gene. The conclusion of the study was that reduced Calcium ATPase might play a role in increased susceptibility to colon cancer.[40]

BREAST CANCER

After skin cancer, breast cancer is the most common cancer for women in America, with one in eight women diagnosed with the disease at some point in their lives.[41] In fact, breast cancer represents 40% of all cancers in women.

Researchers measured Calcium ATPase in normal breast tissue, breast tissue from precancerous lesions, and breast tissue from cancerous lesions in both lobular and ductal breast tissue. In lobular tissue, Calcium ATPase levels were reduced in precancerous lesions and further reduced in cancerous lesions. The same exact pattern appeared in the duct area.

The level of Calcium ATPase loss is different based on each specific stage of breast cancer. For example, Calcium ATPase levels were 50% lower in stage 3 cancer than in stage 1 cancer. Reduced Calcium ATPase levels also correlated with two other tests used to measure breast cancer parameters: the negative hormone receptor expression and the triple negative status in ductal carcinoma. Overall, reduced Calcium ATPase was the most reduced in high nuclear grade (bad), highly proliferating, hormone-receptor negative tumors. Needless to say, the reduction in Calcium ATPase may be involved in the formation of early/premalignant phase of breast cancer development,

39 Brouland JP, Gelebart P, Kovacs T, Enouf J, Grossmann J, Papp B. The loss of sarco/endoplasmic reticulum calcium transport ATPase 3 expression is an early event during the multistep process of colon carcinogenesis. American Journal of Pathology, Vol 167, no 1, July 2005.

40 Korosec B, Glavac D, Rott T, Ravnik-Glavac M. Alterations in the ATP2A2 gene in correlation with colon and lung cancer. Cancer Genetics and Cytogenetics 171(2006) 105–111.

41 http://www.breastcancer.org/symptoms/understand_bc/statistics.

and this knowledge may lead to a new understanding of breast cancer in its early stages.[42]

FUTURE DIRECTIONS

- Although increasing cancer susceptibility and promoting tumor growth, the irony is that reducing Calcium ATPase in certain types of cells can also trigger endoplasmic reticulum stress that leads to cell death. Researchers are working to develop drugs that selectively target cancer cells (smart bombs, so to speak) to induce endoplasmic reticulum stress to promote tumor cell death pathways.[43]

42 Papp B, Brouland J.P. Altered endoplasmic reticulum calcium pump expression during breast tumorigenesis. Breast cancer: Basic and Clinical Research 2011:5.
43 Doan NT, Paulsen ES, Sehgal P, Møller JV, Nissen P, Denmeade SR, Isaacs JT, Dionne CA, Christensen SB. Targeting thapsigargin towards tumors. Steroids. 2015 May;97:2–7. doi: 10.1016/j.steroids.2014.07.009. Epub 2014 Jul 24. PMID: 25065587; PMCID: PMC4696022.

CHAPTER SUMMARY

1. Cancer is marked by abnormal cell differentiation.
2. Noncancerous cells go through a process during which the cells become differentiated or specialized. However, cancer cells have a DNA glitch and do not go through the normal process of differentiation. This enables them to reproduce at an elevated rate.
3. The less the cell differentiation, the worse the cancer.
4. Calcium regulation plays an important role in cell differentiation.
5. In cancer cells, calcium regulation is disturbed.
6. Reduced levels of Calcium ATPase play a role in cancer cells' calcium dysregulation.
7. Reduced Calcium ATPase levels are associated with several forms of cancer including lung, squamous cell, colon, and breast cancer.

CHAPTER SIX

THE BRAIN: NEURODEGENERATION AND NEURODEVELOPMENT

Calcium ATPase plays a critical role in brain function. As in all other cells, the rise and fall of intracellular calcium levels serves as a signal. In the brain, the rise and fall of intracellular calcium levels acts as a trigger to facilitate numerous events, such as neurotransmitter release, learning and memory formation, and neuroplasticity.[44] When Calcium ATPase levels are reduced, intracellular calcium levels become dysregulated in the brain and problems occur. For example, as we age, Calcium ATPase declines, which contributes to age-related brain impairments such as Alzheimer's. At the other end of the spectrum, during the prenatal period, calcium regulation is crucial to neuron growth and neural pathway development. Reduced Calcium ATPase has a negative effect on brain development, which occurs in cases such as fetal alcohol syndrome.

Before we delve deeper into those topics, let's take a more detailed look at the importance of calcium levels in brain function.

44 Strehler E., Plasma membrane Calcium ATPase: Targets of oxidative stress in brain aging and neurodegeneration. World J Biol Chem 2010 September 26; 1(9): 271–280.

CALCIUM AND BRAIN FUNCTION

Intracellular calcium plays a key role in brain function because the levels of intracellular calcium are the primary drivers of synaptic transmissions. Synaptic transmission is the transfer of information via an electrical signal between two neurons. Think of it as how brain cells talk to one another. For neurotransmitters to pass information across synapses, they must be triggered by an electrical signal. This electrical signal then triggers a rise in intracellular calcium. The increase in intracellular calcium causes the release of neurotransmitters, which cross the synapse and bind to receptors on the other side, where the message is delivered. After the message is delivered, calcium levels are returned to baseline. However, if calcium levels remain chronically elevated due to reduced Calcium ATPase levels, more neurotransmitters are released than are needed. These excessive neurotransmitters result in a potentially toxic magnification of the signal or even cell death (termed excitotoxicity). To make matters worse, the neural pathways we use most often are the very pathways most susceptible to the deleterious effects of elevated calcium. This is because these neurons are transmitting information more frequently, creating a greater chance for elevated calcium and damage to both function and viability of these valuable neural pathways.[45]

Elevated intracellular calcium levels caused by reduced Calcium ATPase can also affect the neuron's ability to grow new neurites (those long protrusions you see in images of nerve cells), because it reduces their response to nerve growth factor. Nerve growth factor is responsible for neurite extension. As a result, neurons have fewer and shorter neurites and a significant lack of new growth.[46,47]

Furthermore, reduced Calcium ATPase levels lead to endoplasmic reticulum stress. Endoplasmic reticulum stress can trigger a separate cascade of actions that result in earlier-than-normal programmed cell death, or apoptosis, of the neuron. In other words, you lose brain cells.

45 Toescu E, Verkhratsky A. Parameters of calcium homeostasis in normal neuronal ageing. J. Anat. (2000) 197, pp. 563–569.
46 Brandt PC, Sisken JE, Neve RL, Vanaman TC. Blockade of plasma membrane calcium pumping ATPase isoform I impairs nerve growth factor-induced neurite extension in pheochromocytoma cells. Proc Natl Acad Sci U S A. 1996 Nov 26;93(24):13843–8.
47 Zaidi A. Plasma membrane Calcium ATPases: Targets of oxidative stress in brain aging and neurodegeneration. World J Biol Chem 2010 Sept 26;1(9):271–280.

CALCIUM REGULATION AND THE AGING BRAIN

An understanding of calcium as a key factor in age-related brain dysfunction began to take shape in the 1980s, when a researcher named Zaven Khachaturian first suggested the importance of calcium regulation in the brain. His basic theory was that the brain's ability to regulate intracellular calcium levels declined as a function of age. This decline resulted in small but persistent elevations in brain intracellular calcium levels. Chronic exposure to elevated intracellular calcium ultimately led to neurotoxicity, which, he believed, was responsible for "normal" age-related brain dysfunction. Since his original "calcium hypothesis" was proposed, several studies have supported Khachaturian's theory.

In one study, researchers examined the time it took intracellular calcium levels to return to baseline after a stimulus. Specifically, they reported that it took neurons in the aged rat brain 25 seconds for calcium to return to homeostasis, as compared to 8 seconds for young rat neurons.[48] Additional studies have supported the concept of elevated intracellular calcium levels in the aging brain, reporting that after a stimulus, intracellular calcium levels are 30–85% higher in brain neurons of aged rats as compared to young rats.[49]

Given that intracellular calcium levels are elevated in aged neurons, researchers were interested to see if there was a corresponding decrease in Calcium ATPase levels in the aging brain. To approximate human brains, researchers measured the amount of Calcium ATPase in brain tissue of rats from age five months up to thirty-four months, which would correspond to a normal age span of human beings from infancy to old age. The study showed that Calcium ATPase levels declined progressively with age. From age five months to fourteen months, Calcium ATPase declined 16%, from age fourteen to twenty-two months Calcium ATPase declined an additional 15%, and from age twenty-two to thirty-four months Calcium ATPase declined an additional 16%. Looking at the big picture, Calcium ATPase levels declined 47% over the lifetime of the rat.[50]

---

48 Verkhratsky A, Shmigol A, Kirischuk S, Pronchuk N, Kostyuk P. Age-Dependent Changes in Calcium Currents and Calcium Homeostasis in Mammalian Neurons. Annuals New York Academy of Sciences 1994 Dec 15; 747:365–81.

49 Michaelis M, Foster C, Jayawickreme C. Regulation of calcium levels in brain tissue from adult and aged rats. Mech Ageing Dev 1992 Mar 1; 62(3):291–306.

50 Zaidi A, Gao J, Squier C, Michaelis L. Age-related decrease in brain synaptic membrane Calcium ATPase in F344/BNF1 rats. Neurobiology of Aging Vol 19, No 5, pp. 487–495.

ALZHEIMER'S AND DISRUPTED CALCIUM HOMEOSTASIS

Just as calcium plays a major role in the dysfunction of the aging brain, impaired calcium regulation is thought to play an important role in the neurodegeneration found in Alzheimer's patients. In addition to the normal age-related decline in Calcium ATPase, in Alzheimer's patients, calcium regulation is further impaired by the disease itself. [51,52]

One hallmark of Alzheimer's disease is the buildup of a sticky substance called *amyloid beta-peptide senile plaques,* which occurs in the areas around nerve cells in the brain. Amyloid is a common form of protein in our bodies that normally causes no issues. For reasons that scientists have yet to pinpoint, Alzheimer's disease causes the amyloid protein to divide improperly, creating a beta version that is toxic to nerve cells in the brain. Amyloid beta-peptide has been shown to inhibit Calcium ATPase in neurons. In one study, brain tissue from four neurologically normal adults was exposed to amyloid beta-peptide, and there was an astonishing 35% reduction in Calcium ATPase levels after one hour.[53]

Further studies provide corroborating evidence. Live imaging of neurons in mice that were made to mimic Alzheimer's disease revealed elevated intracellular calcium levels near amyloid beta plaques, indicating that the net result of amyloid beta interaction with synapses is elevated intracellular calcium levels. Since synaptic transmission is tightly controlled by intracellular calcium levels, it may be that amyloid beta-peptide synaptic dysfunction is related to higher-than-normal intracellular calcium levels.[54]

An additional factor connecting Alzheimer's and Calcium ATPase is the association between Alzheimer's and endoplasmic reticulum stress. Reduced Calcium ATPase leads to endoplasmic reticulum stress, which then triggers the "unfolded protein response" (UPR), a cell program that attempts to correct damaged proteins caused by endoplasmic reticulum stress. If unsuccessful within a specific time

51 Mata A., Impairment of the activity of the plasma membrane Calcium ATPase in Alzheimer's disease. Biochem Soc Trans 2011 June; 39(3):819–22.
52 Berrocal M, Marcos D, Sepulveda M, Perez M, Avila J, Mata A. Altered Ca2+ dependence of synaptosomal plasma membrane Calcium ATPase in human brain affected by Alzheimer's disease. FASEB J, 2009 Jun; 23(6):1826–34.
53 Mark R, Hensley K, Butterfield A, Mattson M. Amyloid Beta-Peptide Impairs Ion-Motive ATPase Activities: Evidence for a Role in Loss of Neuronal Ca2+ Homeostasis and Cell Death. The Journal of Neuroscience, September 1995, 15(9): 6239–6249.
54 Green K. Calcium in the initiation, progression and as an effector of Alzheimer's disease pathology. Vol 13, Issue 9a Sept 2009: 2787–2799.

frame, the UPR will trigger cell death or apoptosis. Neuron death is the most devastating aspect in the Alzheimer's disease progression. New molecules that stimulate Calcium ATPase have been shown to reduced endoplasmic reticulum stress and prevent neuron death in Alzheimer's animal models.[55]

NEURODEVELOPMENT AND CALCIUM ATPASE

A seminal research study conducted at Wake Forest University in 2008[56] was the first to examine the effect of disturbed intracellular calcium levels on the development of individual neurons that takes place during prenatal and postnatal neurodevelopment. Researchers exposed developing neurons to a substance called thapsigargin that inhibits Calcium ATPase. The neurons were exposed during periods of time that correlated with the last trimester of pregnancy and first 1–2 years in human development. Those time frames were chosen because they are associated with stage-specific changes in neuronal maturation. The results were staggering. The reduction in Calcium ATPase had profound negative effects on key markers of neuronal development. For example, neurite growth indicators, neurite length, and neurite complexity were all fundamentally reduced at all levels of exposure. Furthermore, even upon removal of the Calcium ATPase inhibitor, it prevented the establishment of neuronal networks that are crucial during a later stage of brain and nervous system development. Thus, both acute effects and long-term outcomes were negatively affected.

The researchers' conclusion was that the maintenance of calcium homeostasis during neurodevelopment is of such importance that avoiding medications and substances that disrupt calcium homeostasis should be considered a preventive strategy to limit brain injury in children.

55 Krajnak K, Dahl R. A new target for Alzheimer's disease: a small molecule SERCA activator is neuroprotective in vitro and improves memory and cognition in APP/PS1 mice. Bioorg Med Chem Lett 2018 May 15;28(9): 1591–1594.
56 Ringler, Sarah L, et al. Effects of disrupting calcium homeostasis on neuronal maturation: early inhibition and later recovery. Cellular and molecular neurobiology vol. 28,3 (2008): 389–409. doi:10.1007/s10571-007-9255-9.

Neurite Length

- During early stages of development, neurons grow rapidly, extending neurites in search of new connections.
- Calcium ATPase inhibition reduces neurite length.

Neurite Complexity

- During early stages of development, neurons increase in complexity.
- Calcium ATPase inhibition reduces neurite complexity.

Neuronal Networks

- During later stages, neuronal networks are solidified.
- Calcium ATPase inhibition prevents neuronal networks from normal development.

FETAL ALCOHOL SYNDROME AND CALCIUM ATPASE

Fetal alcohol syndrome is a condition that results from profound alcohol exposure during the mother's pregnancy. The severity of fetal alcohol syndrome symptoms varies widely. Symptoms can include a small brain and head circumference, heart and kidney problems, low IQ, learning deficits, ADHD, mood dysregulation, and social difficulties. Effects caused by fetal alcohol syndrome are often irreversible.

Researchers used rats to examine the effects that alcohol consumption has on Calcium ATPase during pregnancy. Calcium ATPase levels were significantly reduced in the alcohol-exposed offspring, and the related decrease in neuronal development produced profound alterations in brain function.[57]

Another study supported this finding. It measured Calcium ATPase levels in six areas of the brain at perinatal five, fifteen, and twenty-five days of age after alcohol exposure in utero. Alcohol

57 Guerri C, Esquifino A, Sanchis R, Grisolía S. Growth, enzymes and hormonal changes in offspring of alcohol-fed rats. Ciba Found Symp. 1984;105:85-102. doi: 10.1002/9780470720868.ch6. PMID: 6563994.

exposure decreased enzyme activities in those six brain areas when compared to a control group. The reduction in Calcium ATPase was greater the longer the exposure to alcohol.[58]

In a third study, synaptic membranes were isolated from two- and four-month-old rats that were prenatally or both prenatally and postnatally exposed to alcohol. There was a 22–33% reduction in Calcium ATPase. The researchers concluded that prenatal alcohol exposure delays synaptic development and continued alcohol exposure during lactation may alter the physiochemical structure of synaptic membranes.[59]

The effects of chronic alcohol exposure in utero on Calcium ATPase levels in the heart was also investigated. Pregnant guinea pigs were given drinking water and lab chow containing 2.5% ethanol (the active ingredient in alcoholic drinks) from day thirty to the end of pregnancy. A control group received a normal diet. The one- to three-day-old offspring of ethanol-treated mothers had significantly reduced cardiac Calcium ATPase levels. After three to five months with no additional alcohol exposure, the offspring had nearly normal cardiac Calcium ATPase levels. Moderate alcohol exposure in utero produced fundamental cardiac alterations in the newborn that could be slowly reversed with abstinence from alcohol.

FUTURE DIRECTIONS

- In 2018, Dr. Russel Dahl reported that a small molecule compound that targets and activates Calcium ATPase in brain neurons proved successful in improving both memory and cognition in the mice model of Alzheimer's patients. In addition, neuron death was prevented.[60]
- In a separate study, the same compound also activates Calcium ATPase in neuron cells that ameliorates dyskinesia in the mice model of Parkinson's.[61]

58 Rudeen PK, Guerri C. The effects of alcohol exposure in utero on acetylcholinesterase, Na/K-ATPase and Ca-ATPase activities in six regions of rat brain. Alcohol Alcohol. 1985;20(4):417-25. PMID: 3002403.
59 Guerri C. Synaptic membrane alterations in rats exposed to alcohol. Alcohol Alcohol Suppl. 1987;1:467–72. PMID: 2827696.
60 Dahl R. Krajnak K, Dahl R. A new target for Alzheimer's disease: a small molecule SERCA activator is neuroprotective in vitro and improves memory and cognition in APP/PS1 mice. Bioorg Med Chem Lett 2018 May 15;28(9): 1591–1594.
61 A new target for Parkinson's disease: Small molecule SERCA activator CDN1163 ameliorates dyskinesia in 6-OHDA-lesioned rats. Biorg Med Chem 2017 Jan 1;25(1):53–57.

CHAPTER SUMMARY

1. Intracellular calcium is important in brain function.
2. As we age, intracellular calcium can become dysregulated, in part because Calcium ATPase levels tend to decline.
3. When intracellular calcium levels are dysregulated, three things can occur:
   1) neuronal excitotoxicity
   2) reduced neuron growth
   3) neuron death
4. Dysregulated intracellular calcium levels also play a role in Alzheimer's.
5. Intracellular calcium regulation is fundamental for normal neurodevelopment.
6. When intracellular calcium levels are dysregulated, brains do not develop the neural circuitry required for normal brain development.
7. Preventing brain calcium dysregulation via reducing exposure to Calcium ATPase inhibitors is an action we should consider important at any age.

CHAPTER SEVEN

INFLAMMATION

Calcium ATPase plays a major role in inflammation because calcium levels control mast cell degranulation, which is the event that triggers the release of inflammatory substances such as TNF-alpha from mast cells. Reduced Calcium ATPase magnifies the body's response to inflammation.

Inflammation is the body's way of protecting itself. It originates as a localized reaction to infection, irritation, or injury and produces redness, warmth, swelling, and pain. The reaction serves as a means of eliminating the agent that caused the insult and repairing or mitigating tissue damage. Anytime your body detects something foreign, it begins the process of fighting the invader off and causing inflammation. The foreign substance can be an allergen (an allergic trigger), chemical, or in the case of an autoimmune disease, the body itself. A short burst of inflammation, for perhaps a day or so, is not the problem. Our bodies react, adjust, and then settle back into homeostasis. However, when the foreign invader persists (or the body responds as if there were a foreign invader or injury), the result is chronic, ongoing inflammation that does not go away.

Relatively recent research has pointed the finger at various lifestyle factors that can lead to chronic inflammation, including the consumption of certain types of vegetable oils, a high sugar diet, specific food sensitivities, and the state of obesity itself. And as it turns out, reduced Calcium ATPase, and therefore elevated levels of intracellular calcium, also play a critical role in inflammation.

HOW REDUCED CALCIUM ATPASE ACTIVITY INCREASES INFLAMMATION

Intracellular calcium plays a key role in controlling the body's allergic response through specialized cells called *mast cells*. They can be compared to a fire department having a store of special chemicals that can be used to neutralize harmful chemicals in case a tanker truck or railroad car crashes. Different types of chemical spills require different chemicals to neutralize them. Mast cells serve the purpose of determining which chemicals are needed for which inflammatory fire. They contain storage sacs called secretory granules that store *inflammatory mediators* (those chemicals that put out fires), such as histamine, interleukin, and TNF-alpha. The body produces these chemicals in response to injury and the corresponding inflammation.

Mast cells, like all other cells, are regulated by intracellular calcium levels. When calcium levels rise, mast cells are triggered to release inflammatory substances. This release of inflammatory substances is called *mast cell degranulation*.

Because intracellular calcium plays such an integral role in mast cell degranulation, researchers studied the role Calcium ATPase plays in it, as well. They began with a chemical that inhibits Calcium ATPase activity and determined how it affected the amount of histamine (an inflammatory mediator) released due to mast cell degranulation. When mast cells were pretreated with this Calcium ATPase inhibitor, allergic response increased by 55% as compared to untreated mast cells. In addition, researchers found that treatment of mast cells with the Calcium ATPase inhibitor alone resulted in mast cell degranulation even without a trigger.[62] This means that reduced Calcium ATPase activity can cause our bodies to be inflamed even when there's no need. Also, reduced Calcium ATPase magnifies the body's response to an allergic trigger unnecessarily. To sum up, reduced Calcium ATPase is proinflammatory.

In a similar study, researchers looked at the effect that Calcium ATPase inhibitors had on two additional inflammatory mediators (interleukin 4, IL-4, and monocyte chemoattractant protein 1,

62 Kitajima S, Momma J, Tsuda M, Kurokawa Y, Teshima R, Sawada J. Effects of 2,5-di(tert-butyl)-1,4-hydroquinone on intracellular free Ca2+ levels and histamine secretion in RBL-2H3 cells. Inflamm Res 1995 Aug; 44(8):335–39.

MCP-1). They found the same results as for histamine: Reduced Calcium ATPase activity prompted the release of these inflammatory mediators without an allergic trigger.[63,64]

Researchers also examined the effect Calcium ATPase inhibitors had on the release of TNF-alpha, a fourth key inflammatory substance. They reported, again, that even without an allergic trigger, reduced Calcium ATPase caused mast cell degranulation and the release of the inflammatory mediator TNF-alpha.[65]

Another study looked at how much and which type of inflammatory mediators were released from cells when Calcium ATPase was inhibited, compared to how much and which types were released when the cells were exposed to an antigen. They found that mast cells released the same types of inflammatory mediators under both test conditions. They also found that several additional mediators were released during just the inhibition of Calcium ATPase. This means reduced Calcium ATPase activity not only causes mast cells to release inflammatory mediators similar to an allergic response, but that lower Calcium ATPase activity can also cause the release of different, additional mediators that aren't normally released in the presence of that particular antigen.[66]

The most powerful point these studies emphasize is that reduced Calcium ATPase activity can result in inflammation without any trigger present. In addition, reduced Calcium ATPase activity also magnifies allergic responses, thus making allergic reactions more severe. Whichever inflammation you might have going on normally, having reduced Calcium ATPase levels is going to make it a lot worse.

THE VICIOUS INFLAMMATORY CYCLE

To make matters worse, the inflammatory mediators triggered by reduced Calcium ATPase in turn reduce Calcium ATPase in other, unrelated parts of the body. This reveals a powerful link between

63 Onose J, Teshima R, Sawada J. Calcium ATPase inhibitor induces IL-4 and MCP-1 production in RBL-2H3 cells. Immunol Lett 1998 Nov; 64(1):17–22.

64 Teshima R, Onose J, Okunuki H, Sawada J. Effect of Ca(2+) ATPase inhibitors on MCP-1 release from bone marrow-derived mast cells and the involvement of p38 MAP kinase activation. Int Arch Allergy Immunol 2000 Jan; 121(1):34–43.

65 Teshima R, Onose J, Ikebuchi H, Sawada J. Calcium ATPase inhibitors and PKC activation synergistically stimulate TNF-alpha production in RBL-2H3 cells. Inflamm Res 1998 Aug; 47(8):328–33.

66 Nakamura R, Ishida S, Ozawa S, Saito Y, Okunuki H, Teshima R, Sawada J. Gene expression profiling of Ca2atpase inhibitor DTBHQ and antigen-stimulated RBL-2H3 mast cells. Inflamm Res. 2002 Dec;51(12):611–18.

inflammation and chronic diseases. As we saw above, reduced Calcium ATPase activity results in the release of inflammatory mediators. The vicious cycle occurs because these inflammatory mediators, now in the bloodstream, actually reduce Calcium ATPase in various organs throughout the body (similar to the way that high blood sugar reduces Calcium ATPase in numerous parts of the body).

For example, high levels of inflammatory mediators in the bloodstream are directly correlated with reduced Calcium ATPase levels in the heart muscle. You will learn in the following pages that a reduction in Calcium ATPase in the heart muscle is the primary cause of heart failure.

Inflammatory mediators in the bloodstream can also reduce Calcium ATPase in pancreatic cells, which, as we learned in the Diabetes chapter, disrupts insulin production from the cells and is a cause of diabetes.

Another example of the connection between inflammation and Calcium ATPase is inflammatory bowel disease (IBD). In addition to the increase in inflammatory markers in IBD, Calcium ATPase levels in the colon are reduced. Lower Calcium ATPase levels prevent normal colon contractions, leading to poor bowel movements.[67]

67 Al-Jarallah A, Oriowo MA, Khan I. Mechanism of reduced colonic contractility in experimental colitis: role of sarcoplasmic reticulum pump isoform-2. Mol Cell Biochem. 2007 Apr;298(1-2):169–78. doi: 10.1007/s11010-006-9363-8. Epub 2006 Nov 25. PMID: 17131044.

CHAPTER SUMMARY

1. Inflammation is a multifaceted phenomenon associated with numerous chronic diseases.
2. Mast cells play a key role in inflammation.
3. In an allergic response, mast cells release inflammatory substances such as histamine, interleukin, and TNF-alpha. This is known as mast cell degranulation.
4. Reduced Calcium ATPase triggers mast cell degranulation that mimics an allergic response, even if there is no allergen present.
5. Reduced Calcium ATPase actually magnifies the body's response to allergens, creating unnecessary inflammation.
6. Inflammation can lead to reduced Calcium ATPase throughout the body, particularly in the heart muscle (contributing to cardiovascular disease) and the pancreas (contributing to the development of diabetes).

CHAPTER EIGHT

HIGH BLOOD PRESSURE AND ATHEROSCLEROSIS

Calcium ATPase plays a role in both high blood pressure and atherosclerosis. I have chosen to cover high blood pressure and atherosclerosis in the same chapter because high blood pressure is often a precursor to atherosclerosis. In addition, high blood pressure and atherosclerosis often set the stage for further cardiovascular disease, such as heart attacks and atrial fibrillation.

BLOOD PRESSURE

What exactly is blood pressure? It is simply the force, or pressure, of the blood against the walls of the arteries as that blood is circulated throughout the body.

You may not be aware that arteries are made up primarily of muscle, and as we now know, Calcium ATPase plays a primary role in muscle function. The muscle in our arteries is similar to the muscles in our arms; like a bicep, arteries contract and relax. However, unlike biceps, arteries are not under our voluntary control. Rather, the contraction and relaxation of the arteries is involuntary and controlled by specific mechanisms within the body. Because their function is involuntary, they are in the "smooth muscle" category.

HIGH BLOOD PRESSURE AND CALCIUM ATPASE

As in all cells, a rise in intracellular calcium levels sends a signal to trigger an event. In the case of arteries, a rise in intracellular calcium triggers the contraction of the artery smooth muscle, called vasoconstriction. Following the contraction, intracellular calcium levels are returned to resting levels by Calcium ATPase. The reduced intracellular calcium levels cause the muscle to relax, called vasodilation.

Maintaining optimal blood pressure levels is a finely tuned process. Too much vasoconstriction will result in high blood pressure, and too much vasodilation will result in low blood pressure (which is not usually as much of a problem as high blood pressure but can be in extreme cases). How does the body regulate blood pressure? The body regulates blood pressure by releasing a substance called nitric oxide from the inner lining of the artery (called the endothelium) when blood pressure is too high. Nitric oxide in turn stimulates Calcium ATPase to reduce the elevated intracellular calcium levels in the artery smooth muscle. When intracellular calcium levels are lowered, the artery muscle relaxes, which reduces blood pressure.

Researchers measured Calcium ATPase levels in fifty-five patients with high blood pressure and compared them to thirty-three patients with normal blood pressure. The results showed that the high blood pressure group had significantly lower Calcium ATPase levels as compared to the control group. No surprise, then, that the high blood pressure group also had significantly higher intracellular calcium levels. The net effect was increased vascular tone (the degree of contraction maintained by the artery) resulting in higher blood pressure.[68] Other studies have confirmed the connection between low Calcium ATPase levels and high blood pressure.[69,70]

In another study, researchers looked at ninety-two patients: forty-five with high blood pressure, nine with borderline high blood pressure, and thirty-eight with normal blood pressure. The differences in platelet intracellular calcium levels were significant.

---

68 Fu Y, Wang S, Lu Z, Li H, Li S. Erythrocyte and plasma Calcium, Mg2+ and cell membrane adenosine triphosphate activity in patients with essential hypertension. Chin Med J 1998 Feb; 111(2):147–9.

69 Mizuno H, Ikeda M, Harada M, Onda T, Tomita T. Sustained contraction to angiotensin II and impaired Calcium-sequestration in the smooth muscle of stroke-prone spontaneously hypertensive rats. Am J Hypertens 1999 Jun; 12(6):590–5.

70 Shi B. Clinical study on the relationship between erythrocyte ATPase activity and lipid peroxidation in essential hypertension. Zhonghua Xin Xue Guan Bing Za Zhi 1993 Feb; 21(1):26–28, 63.

Borderline patients had 15% higher intracellular calcium levels as compared to those in the control group, whereas patients with high blood pressure had intracellular calcium levels 55% greater than those in the control group.

The researchers took it one step further and measured intracellular calcium levels after the high blood pressure patients were treated with one of three heavily prescribed blood pressure medications (calcium channel blockers, ACE inhibitors, or beta blockers). The medications not only reduced the high blood pressure, but also lowered intracellular calcium to normal levels, providing further evidence of the connection between intracellular calcium levels and blood pressure.[71]

An important fact to remember is that blood pressure sets the stage for atherosclerosis, which leads to heart attack and stroke. We go through this process in more detail in the next section, on atherosclerosis.

ATHEROSCLEROSIS

You have probably heard the term *atherosclerosis* before, but to be clear, let's look at a clinical definition. Atherosclerosis is defined as the hardening and narrowing of arteries (blood vessels that carry oxygen from the heart to other parts of the body) due to a buildup of plaque.

Plaque is primarily made up of bad cholesterol and fats, as well as free-floating calcium in the bloodstream that accumulates on the interior of the arteries (imagine a buildup of sludge in household pipes).[72]

How and why does plaque form? The first step in plaque formation is an injury to the inner lining of the artery called the endothelium. This injury can happen from a wide variety of factors, such as smoking, diabetes, and high blood pressure. High blood pressure by its very nature is associated with damage to the endothelium layer due to the wear and tear caused by the increased pressure of the blood against the artery walls.

LDL cholesterol, calcium, and other substances latch onto the injuries in the artery wall and form plaque, restricting blood flow to

71 Paul E, Bolli P, Burgisser E, Buhler F. Correlation of platelet calcium with blood pressure; effect of antihypertensive therapy. New England Journal of Medicine. April 26, 1984.
72 https://www.ncbi.nlm.nih.gov/pubmedhealth/PMHT0023229/.

many parts of the body, including the brain, heart, kidneys, arms, and legs.[73] The body responds to this injury just like it responds to a scab on your knee; it tries to repair it.[74] In response to the injury, damaged endothelial cells release substances called *cytokines*. Cytokines are molecules whose job it is to orchestrate an immune response to the injury by stimulating the movement of various types of cells toward sites of inflammation, infection, and trauma.

One of the responses triggered by the release of cytokines is a signal for smooth muscle cells (which make up the outer layer of the artery) in the region to migrate toward the site of the injury in the endothelial layer and plug up the hole, so to speak. Normally this is good, but if left unchecked, this cell migration causes problems. When smooth muscle cells are where they belong at the outer layer of the artery, they perform the function of providing flexibility to the vessel, allowing it to expand and contract in response to the additional pressure caused by each beat of the heart. If they migrate from the outer layers of the artery wall into the endothelium, where they don't belong, it can lead to the formation of a fibrous capsule covering the plaque that has collected at the site of the injury. This solidifies the plaque, making it less flexible.[75] The combination of a buildup in plaque and reduced flexibility in the arterial wall can cause a rupture and a heart attack or stroke to occur. (I go into more detail about heart attacks and strokes in the next chapter.)

The body's natural mechanism for balancing excessive cell migration is to release nitric oxide, a molecule manufactured in the endothelium, to offset the cytokines' stimulation of cell migration.[76] Nitric oxide acts to reduce cell migration through the activation of Calcium ATPase, which reduces the elevation of intracellular calcium. Therefore, Calcium ATPase helps to prevent excessive cell migration, reduces the occurrence of atherosclerosis, and thereby reduces the likelihood of heart attacks and strokes.

73 https://www.livestrong.com/article/414110-how-to-reverse-athersclerosis-with-diet/.
74 Nakao J, Ito H, Ooyama T, Chang W, Murota S. Calcium dependency of Aortic Smooth muscle cell migration induced by 12-l-Hydroxy-5,8,10,14-eicosatetraenoic Acid. Athero 46 (1983) 309–19.
75 Kraemer R. Regulation of cell migration in atherosclerosis. Curr Athero Rep 2000 Sep; 2(5):445–52.
76 Ying J, Tong X, Pimentel D, Weisbrod R, Trucillo M, Adachi T, Cohen R. Cysteine-674 of the Sarco/Endoplasmic reticulum calcium ATPase is required for the inhibition of cell migration by nitric oxide.Arterioscler Thromb Vasc. Biol Jan 2007.

When Calcium ATPase activity is reduced, like with diabetes, nitric oxide is not able to do its job and excess cell migration can occur, which contributes to atherosclerosis in diabetics.[77] Further evidence of the role Calcium ATPase plays is a recent study that demonstrated that a Calcium ATPase gene transfer (that increases Calcium ATPase) into a rat artery reduces smooth muscle cell migration to the endothelial layer after injury.[78]

77 Tong X, Ying J, Pimentel DR, Trucillo M, Adachi T, Cohen RA. High glucose oxidizes SERCA cysteine-674 and prevents inhibition by nitric oxide of smooth muscle cell migration. J Mol Cell Cardiol. 2008 Feb; 44(2):361–69.
78 Lipskaia L, Hadri L, Lopez J, Hajjar R, Bobe R. Benefit of SERCA2a Gene transfer to vascular endothelial and smooth muscle cells: a new aspect in therapy of cardiovascular diseases. Current Vascular Pharmo 2013,11,465–79.

CHAPTER SUMMARY

1. Blood pressure is simply the force of the blood, against the artery walls.
2. Blood pressure is essential to pump blood as it delivers oxygen and removes toxins throughout the body. High blood pressure occurs when the arteries become too narrowed due to excess muscle constriction. The excess muscle constriction is caused by elevated intracellular calcium.
3. The body modulates blood pressure through the release of nitric oxide. Nitric oxide stimulates Calcium ATPase, which reduces elevated intracellular calcium and lowers blood pressure.
4. High blood pressure damages the lining of the artery wall, which sets the stage for atherosclerosis.
5. Atherosclerosis is a condition where arteries get clogged by plaque that is made up of LDL cholesterol and other substances.
6. The first step in the formation of plaque is an injury to the artery wall.
7. At the site of the injury, cholesterol accumulates.
8. In response to the injury, the body triggers "cell migration."
9. Cell migration leads to hard capsules forming over the plaque. These hard capsules are more damaging if they enter the bloodstream because they can block blood flow to the heart and brain.
10. The body's mechanism for preventing excess cell migration is the release of nitric oxide from the endothelium.
11. Nitric oxide utilizes Calcium ATPase to reduce cell migration.

CHAPTER NINE

HEART ATTACKS AND STROKES

Calcium ATPase plays an important role in heart attacks and strokes. As we covered in Chapter Eight, reduced Calcium ATPase contributes to high blood pressure and atherosclerosis, setting the stage for heart attacks and strokes. Reduced Calcium ATPase also increases intracellular calcium in blood platelets, contributing to the formation of blood clots that can lead to heart attacks and strokes.

HEART ATTACKS

A heart attack occurs when plaque on the artery wall ruptures and blood clots are formed, completely blocking passage of blood within the artery to the heart. This can occur because of a continual buildup of plaque in the arterial wall, which eventually, given the constant pressure of blood flow, ruptures and damages the surrounding tissue. The body responds by rushing clotting factors to the rupture, and blood clots are formed. The combination of the plaque and the blood clots can then completely block the flow of oxygenated blood to the heart, causing damage to the heart muscle in the area where the artery was supplying blood.[79,80] Once the heart is damaged, it is more susceptible to heart failure, which we will discuss in the next chapter.

79 https://www.heart.org/en/health-topics/heart-attack/about-heart-attacks.
80 https://www.webmd.com/heart-disease/heart-disease-heart-attacks#1.

STROKE

The two major arteries on each side of your neck that provide blood flow to your brain are called the *carotid arteries*. When plaque builds up in the carotid arteries, oxygenated blood flow to the brain is restricted. If blood to the brain is restricted temporarily, the individual may have a *transient ischemic attack* (TIA), also called a ministroke. Symptoms can include confusion, vision loss, slurred speech, difficulty speaking, weakness, numbness, and even paralysis.

If blood flow to the brain is completely blocked, it is called a stroke. A stroke differs from a TIA in that oxygenated blood is blocked from a portion of the brain for an extended period of time, and as a result that brain tissue dies. Depending on which part of the brain is affected, this can cause severe impairment to movement, memory, speech, or vision. Approximately 87% of strokes are related to atherosclerosis, which are termed *ischemic strokes*.[81]

CALCIUM ATPASE AND BLOOD CLOT FORMATION

Calcium ATPase plays a role in both high blood pressure and atherosclerosis, which in turn set the stage for heart attack and stroke. In this section, we are going to discuss the additional role Calcium ATPase plays in heart attacks and strokes via its impact on blood clot formation, a primary event in heart attack and stroke.

As we explained above, blood clots, which are triggered by loosening plaque, play a key role in both heart attacks and strokes because they block the passage of oxygenated blood. Blood clots form through the process of *platelet aggregation*. Platelet aggregation, as the name implies, is the joining together of blood platelets. When joined together, blood platelets form clumps, which result in the clotting of blood. Platelet aggregation plays a life-saving role in the body. When you accidentally cut yourself while slicing onions, platelet aggregation is what slows and then stops the bleeding; without it, a simple cut could cause you to bleed to death.

In platelets, like in all other cells in the body, an increase in intracellular calcium triggers a response, in this case platelet aggregation. *Thrombin*, which is an enzyme in the blood and a major factor in

81 https://www.stroke.org/en/about-stroke/types-of-stroke/ischemic-stroke-clots.

blood clotting, is released after an injury (e.g., rupture of plaque in artery) and triggers platelet aggregation by causing an increase in intracellular calcium within platelets. This increased intracellular calcium stimulates platelet aggregation, and thus the formation of blood clots. Without the increase in intracellular calcium, our blood would not clot.

In response to platelet aggregation, a molecule called nitric oxide (which comes in handy in a lot of situations!) is released by the endothelium cells in the artery wall as a counterbalance to thrombin, thus returning platelet stickiness to optimal levels. Nitric oxide works by stimulating Calcium ATPase. Calcium ATPase lowers intracellular calcium levels, thereby reducing platelet aggregation.

Let's look at research regarding nitric oxide, Calcium ATPase, and platelet aggregation.

In one study, researchers measured the effectiveness of nitric oxide on platelet aggregation when Calcium ATPase activity was blocked.[82] In other words, how effective was nitric oxide in preventing blood clotting without the help of Calcium ATPase? Using blood samples from healthy volunteers, platelets were isolated via a centrifuge to create platelet-rich plasma. The study reported that when platelets were pretreated with a Calcium ATPase inhibitor, nitric oxide's ability to reduce platelet aggregation was eliminated. Why? Because that ability hinges on the reduction of intracellular calcium, and Calcium ATPase is required to do that. What this study clearly demonstrates is the importance of Calcium ATPase to modulate platelet aggregation, which reduces the likelihood of blood clot formation associated with heart attack and stroke.

So pulling it all together, when it comes to heart attacks and strokes, reduced Calcium ATPase plays a key role in high blood pressure and atherosclerosis, as well as the increased likelihood of blood clot formation.

82 Trepakova ES, Cohen RA, Bolotina VM. Nitric Oxide Inhibits Capacitative Cation Influx in Human Platelets by Promoting Sarcoplasmic/Endoplasmic Reticulum Ca 2+-ATPase–Dependent Refilling of Calcium Stores Circ Res. 1999; 84:201–209.

CHAPTER SUMMARY

1. The primary factor underlying heart attacks and stroke is atherosclerosis. Atherosclerosis means that plaque has formed on the artery walls like sludge in pipes. This plaque reduces blood flow.
2. Heart attacks are associated with clogged arteries to the heart.
3. Strokes are associated with clogged arteries to the brain.
4. In both cases, ruptured plaque leads to blood clot formation. Blood clots block oxygenated blood flow to the heart or brain. Without oxygen, parts of the heart or brain are damaged.
5. Calcium ATPase also plays a key role in reducing the formation of blood clots through modulation of platelet aggregation.

CHAPTER TEN

HEART FAILURE AND ATRIAL FIBRILLATION

In the last chapter, we covered heart attacks, and now we move on to heart failure and atrial fibrillation. Heart failure is different from a heart attack. Heart attacks have to do with blood flow to the heart. Heart failure is related to the heart muscle itself. Atrial fibrillation is related to the rhythm of the heartbeat.

Calcium ATPase is a primary factor in heart failure because the heart is a muscle that contracts and relaxes, and Calcium ATPase is essential to that process. Atrial fibrillation is related to Calcium ATPase because intracellular calcium levels play an essential role in maintaining a normal heartbeat.

HEART FAILURE

The most basic definition of heart failure is that the heart is not pumping sufficiently. The net result is that inadequate oxygen and nutrients are delivered to cells and, hence, every organ in the body. The lack of oxygen and nutrients can manifest itself in numerous ways, such as shortness of breath, fatigue, and swollen hands and feet. In addition, subtler signs may manifest, like cognitive impairment due to lack of oxygen to the brain. Studies have shown that heart failure patients are more than four times as likely to experience cognitive impairment than control subjects.[83]

In the early stages of heart failure, the body has a few tricks up its sleeve to compensate for the heart's reduced output of oxygenated blood. The heart enlarges so it can contract more strongly and keep up with the demand to pump more blood. It begins to pump faster, creating an increased heart rate, and blood vessels narrow to create more pressure to push blood through the weakened heart. These adjustments work in the short term but over time are unable to stop the progression of heart failure. Unfortunately, there is no cure.

HEART FUNCTION AND CALCIUM ATPASE

As discussed above, it's important to remember that the heart is a muscle. The muscle needs to contract to pump blood to the body, but equally important, the muscle needs the ability to relax in order for blood to fill its atria and ventricles before the next beat. Both contraction and relaxation are crucial for the heart to pump efficiently. Intracellular calcium levels play a fundamental role in this process.

In order for the heart to contract, the level of calcium within the cells needs to rise. In order for the heart to relax, the level of calcium within the cells needs to fall. Calcium ATPase is the worker bee responsible for making this process happen. After the heart contracts, Calcium ATPase clears the excess calcium out of the heart muscle cells, enabling it to relax.

WHEN THERE'S NOT ENOUGH CALCIUM ATPASE

For decades, reduced levels of Calcium ATPase have been reported in both animal and human end-stage heart-failure patients.

For example, in 2001, researchers examined the relationship between heart failure and reduced Calcium ATPase. They secured twenty-one donor hearts that could not be used for transplantation. Of these twenty-one hearts, eleven were from individuals

83 http://www.heart.org/HEARTORG/.

with no evidence of cardiac disease, and ten were from individuals with enlarged hearts but with normal cardiac function. These hearts were then compared with nine failing hearts from transplant recipients. The failing hearts had an approximately 58% reduction in Calcium ATPase enzyme levels as compared to normal hearts with no cardiac disease. Hearts that were enlarged but not failing had a 20% reduction in Calcium ATPase enzyme levels as compared to normal hearts with no cardiac disease.[84]

In 2003, researchers looked at things from a different angle to determine if lower Calcium ATPase levels might speed up the development of heart failure. To test this, they bred genetically manipulated mice with an altered Calcium ATPase gene, which resulted in approximately 20% lower Calcium ATPase levels. The mice were subjected to ten weeks of pressure overload to mimic cardiovascular stress conditions. At the end of ten weeks, 64% of mice with reduced Calcium ATPase levels were in heart failure as compared to 0% in mice without the altered Calcium ATPase gene.[85] What this tells us is that reduced Calcium ATPase levels may make you more susceptible to heart failure if you are under cardiac stress conditions, such as what happens when an artery is clogged.

Numerous other studies have confirmed the association between decreased Calcium ATPase and heart failure and shown that the ensuing calcium dysregulation is a key factor in cardiac dysfunction.[86]

ATRIAL FIBRILLATION

A cousin to heart failure is atrial fibrillation, often called AFib or AF, and is the most common type of heart arrhythmia. Arrhythmia is when the heart beats out of rhythm—either too slowly, too quickly, or irregularly. AFib can occur in brief episodes, or it can be a permanent condition. AFib increases susceptibility to heart attack, stroke, and heart failure.[87]

84 DiPaola NR, Sweet WE, Stull LB, Francis GS, Schomisch Moravec C. Beta-adrenergic receptors and calcium cycling proteins in non-failing, hypertrophied and failing human hearts: transition from hypertrophy to failure. J Mol Cell Cardiol. 2001 Jun; 33(6):1283–95.

85 Schultz J, Glascock B, Witt S, Nieman M, Nattamai K, Liu L, Lorenz J, Shull G, Kimball T, Perasamy M. Accelerated onset of heart failure in mice during pressure overload with chronically decreased SERCA2 calcium pump activity. Am J Physiol Heart Circ Physiol 286: H1146-H1153, 2004.

86 Zarain-Herzberg A. Regulation of the sarcoplasmic reticulum Calcium-ATPase expression in the hypertrophic and failing heart. Can J Physiol Pharmacol. 2006 May; 84(5):509–21.

87 Lugenbiel P, Wenz F, Govorov K, Schweizer P, Katus H, Thomas D. Atrial fibrillation complicated by heart failure induces distinct remodeling of calcium cycling proteins. PLos One 2015 Mar 16;1093).

One key factor in the association between AFib and heart failure is Calcium ATPase. AFib causes a process called heart restructuring. The heart tries to adapt to the malfunction by getting bigger and by changing cellular and intracellular elements, including a reduction in Calcium ATPase levels. These reduced levels of Calcium ATPase in turn contribute to or hasten heart failure because Calcium ATPase is crucial to the heart muscle's function.

A 2012 study examined just this process. Researchers took rat cardiac cells and applied electrical field stimulation to mimic AFib. After making the cells pace irregularly for twenty-four hours, there was a significant reduction in Calcium ATPase, which was accompanied by a 59% reduction in the strength of heart contractions.[88] Other studies have confirmed reduced levels of Calcium ATPase in patients with AFib.[89]

---

88 Ling L, Khammy O, Byrne M, Amirahmadi F, Foster A, Li G, Zhang L, dos Remedios C, Chen C, Kaye D. Irregular rhythm adversely influences calcium handling in ventricular myocardium: implications for the interaction between heart failure and atrial fibrillation. Circ Heart Fail. 2012 Nov; 5(6):786–93.

89 Ling-Ping Lai, Ming-Jai Su, Jiunn-Lee Lin, Fang-Yue Lin, Chang-Her Tsai, Yih-Sharng Chen, Shoei K, Huang S, Yung-Zu Tseng, Wen-Pin Lien. Down-regulation of L-type calcium channel and sarcoplasmic reticular Calcium ATPase mRNA in human atrial fibrillation without significant change in the mRNA of ryanodine receptor, calsequestrin and phosphlamban. An insight into the mechanism of atrial electrical remodeling. J Am Coll Cardiol 1999; 33:1231–1237.

FUTURE DIRECTIONS

- Three separate studies conducted by researchers at Prassis Sigma-Tau Research Institute in Italy, Cambridge University in England, and Northwestern University in the United States demonstrated that *Istaroxime,* a recently approved cardiac drug, stimulates Calcium ATPase in the heart. This in turn promotes normal heart function in patients with heart failure.[90,91,92]
- In addition, researchers at the Yamaguchi University School of Medicine in Japan recently identified a separate compound that stimulates Calcium ATPase in human cardiac muscle cells and improves heart function in rat studies.[93]
- Research involving gene therapy to increase the amount of Ca2+ATPase in various tissues has met with laboratory success in animal studies. In animal studies focused on heart failure and atherosclerosis, gene therapy has proven to be a potentially viable pathway to address disease states associated with intracellular calcium dysregulation.[94,95,96] However, the progress in human studies has been mixed, but ongoing research continues.

---

90 Huang CL. SERCA2a stimulation by istaroxime: a novel mechanism of action with translational implications. Br J Pharmacol. 2013 Oct;170(3):486–8. doi: 10.1111/bph.12288. PMID: 23822610; PMCID: PMC3791988.

91 Ferrandi M, Barassi P, Tadini-Buoninsegni F, Bartolommei G, Molinari I, Tripodi MG, Reina C, Moncelli MR, Bianchi G, Ferrari P. Istaroxime stimulates SERCA2a and accelerates calcium cycling in heart failure by relieving phospholamban inhibition. Br J Pharmacol. 2013 Aug;169(8):1849–61. doi: 10.1111/bph.12278. PMID: 23763364; PMCID: PMC3753840.

92 Khan H, Metra M, Blair JE, Vogel M, Harinstein ME, Filippatos GS, Sabbah HN, Porchet H, Valentini G, Gheorghiade M. Istaroxime, a first in class new chemical entity exhibiting SERCA-2 activation and Na-K-ATPase inhibition: a new promising treatment for acute heart failure syndromes? Heart Fail Rev. 2009 Dec;14(4):277–87. doi: 10.1007/s10741-009-9136-z. Epub 2009 Feb 24. PMID: 19238540.

93 Kaneko M, Yamamoto H, Sakai H, Kamada Y, Tanaka T, Fujiwara S, Yamamoto S, Takahagi H, Igawa H, Kasai S, Noda M, Inui M, Nishimoto T. A pyridone derivative activates SERCA2a by attenuating the inhibitory effect of phospholamban. Eur J Pharmacol. 2017 Nov 5;814:1–8. doi: 10.1016/j.ejphar.2017.07.035. Epub 2017 Jul 20. PMID: 28734932.

94 Hayward C, Banner NR, Morley-Smith A, Lyon AR, Harding SE. The Current and Future Landscape of SERCA Gene Therapy for Heart Failure: A Clinical Perspective.Hum Gene Ther. 2015 May;26(5):293–304.

95 Lipskaia L, Bobe R, Chen J, Turnbull IC, Lopez JJ, Merlet E, Jeong D, Karakikes I, Ross AS, Liang L, Mougenot N, Atassi F, Lompré AM, Tarzami ST, Kovacic JC, Kranias E, Hajjar RJ, Hadri L. Synergistic role of protein phosphatase inhibitor 1 and sarco/endoplasmic reticulum Ca2+ -ATPase in the acquisition of the contractile phenotype of arterial smooth muscle cells. Circulation. 2014 Feb 18;129(7):773–85.

96 Goonasekera SA, Lam CK, Millay DP, Sargent MA, Hajjar RJ, Kranias EG, Molkentin JD. Mitigation of muscular dystrophy in mice by SERCA overexpression inskeletal muscle. J Clin Invest. 2011 Mar;121(3):1044–52.

CHAPTER SUMMARY

1. The heart is a muscle.
2. In order for the heart to function properly, it has to relax and contract.
3. Intracellular calcium levels regulate this relaxation and contraction of the heart muscle.
4. Calcium ATPase is the primary method of regulating intracellular calcium levels in the heart. Reduced levels of Calcium ATPase are associated with heart failure and atrial fibrillation.

CHAPTER ELEVEN

MUSCLE STRENGTH

There are approximately 640 different skeletal muscles throughout the human body.

Skeletal muscles are those muscles attached to our bones that support our entire skeletal system, which enables us to function normally throughout the day. Without them, we would not be able to move our bones, meaning no walking, sitting up straight, or even lifting a pencil. Most of you are quite familiar with these muscles when you visit the gym: biceps, triceps, quads, pecs, lats, abs, and the like. Calcium ATPase plays an important role in both the contraction and relaxation of all skeletal muscles.

MUSCLE CONTRACTION AND CALCIUM ATPASE

While cardiac and smooth muscles are under involuntary control (for example, we're not in charge of our stomach churning or our heart beating), skeletal muscles are under voluntary control. We need to want to move our leg before we do so. Thus, the impetus for a skeletal muscle contraction begins in the brain. An electrical signal is transmitted from the brain to the appropriate muscles through special neurons called *motor neurons*. Much like an electrical current, brain signals travel down our nerves to the muscles via a series of motor neurons.

The multistep process takes place when an electrical signal reaches the muscle (termed the *motor endplate*) and must cross the neuromuscular junction (the gap between the neurons and the muscles). To accomplish this, a chemical neurotransmitter called *acetylcholine* is released from the endplate into the space between the neuron and muscle. Acetylcholine crosses the gap and binds to specific acetylcholine receptors on the muscle. These receptors then transmit the electrical signals to the muscle, causing the release of calcium from the sarcoplasmic reticulum inside the muscle cells. The end result is an increase in intracellular calcium levels, causing the muscles to contract.

Following contraction, the muscle needs to relax. Once a contraction is completed, the brain stops sending its electrical signal to the muscle, and therefore no more calcium is released from the sarcoplasmic reticulum. Calcium ATPase pumps intracellular calcium back into the sarcoplasmic reticulum, lowering intracellular calcium and enabling muscles to relax. Furthermore, by refilling the sarcoplasmic reticulum with calcium, the muscle is ready to receive the next electrical signal. This entire process happens in under a second and repeats millions of times in a day.

MUSCLE FATIGUE AND CALCIUM ATPASE

We have all experienced muscle fatigue. You've walked or run miles and just can't go any farther. Or maybe you're in a gym and can't force even one more biceps curl. A more formal definition of muscle fatigue is a sustained reduction in muscle force after repetitive contractions. You do ten biceps curls and are beginning to feel fatigue, so on the eleventh curl you may not lift with as much force. After doing a few more? You may not even be able to complete the rep. However, muscle fatigue also occurs in our everyday life and can be a debilitating symptom of clinical conditions such as congestive heart failure and chronic fatigue.

Muscle fatigue is associated with reduced muscle contraction (losing strength), so it's clear that Calcium ATPase plays a role, as it's integral to the contraction and relaxation of muscles. As we discussed above, the final step in muscle contraction is a rise in intracellular calcium, resulting from the release of calcium from the sarcoplasmic reticulum. We also know that Calcium ATPase is responsible for maintaining the sarcoplasmic reticulum calcium stores. Therefore, without sufficient quantities of Calcium ATPase, there will not be adequate calcium available in the sarcoplasmic reticulum to trigger muscle contractions.

Researchers set out to investigate the relationship between Calcium ATPase and muscle fatigue. Eight active but untrained healthy females (average age of twenty-one) were asked to perform knee extensions for thirty minutes at 60% of their maximum load, with a five-second contraction/five-second relaxation cycle. Measurements of maximal contraction (the heaviest weight at which a rep can be done) were taken at rest and again after thirty minutes of exercise. Tissue samples were also taken via needle biopsy to measure Calcium ATPase over these same time frames. After thirty minutes of exercise, Calcium ATPase levels declined by approximately 33%, which paralleled a decline in maximal contraction (the heaviest weight they could lift) of 25%. The researchers concluded that Calcium ATPase levels were reduced after exercise and that this reduction was correlated with a reduction in maximal contraction.[97]

MUSCLE RELAXATION AND CALCIUM ATPASE

As vital as muscle contraction is to our daily lives, so too is its equal and opposite function, muscle relaxation. Without the ability of our muscles to relax after contraction, we would not be able to walk, talk, eat—you name it. Calcium ATPase also plays a critical role in the process of muscle relaxation.

As we discussed previously, intracellular calcium levels are elevated during a muscle contraction. In order for the muscle to relax, intracellular calcium levels must be brought back to resting levels, and Calcium ATPase is the mechanism that does this. This is especially important during exercise, where repetitive muscle contractions occur, and relaxation is important for the muscle to contract again efficiently.

97 Tupling R, Green H, Grant S, Burnett M, Ranney D. Postcontractile force depression in humans is associated with an impairment in SR Ca2+ pump function. Am. J. Physiol Regulatory Integrative Comp. Physiol 278:R87-92, 2000.

Researchers examined the relationship between Calcium ATPase and the ability of the muscle to relax after exercise. As we discussed above, when muscles are fatigued, they are unable to contract as strongly. The flip side is: when muscles are fatigued, they also have a reduced ability to relax. Five subjects (average age of twenty-six) performed a leg-extension test with a weight selected to produce fatigue at approximately three minutes. Prior to and immediately after the test, subjects performed isometric tests to determine muscle relaxation time. Muscle biopsies were also obtained.

Results indicated that it took much longer for muscles to relax after exercise than before exercise. In fact, it took 200% longer, from 62 milliseconds preexercise as compared to 149 milliseconds postexercise (measured by what is called half-muscle relaxation time). In concert, Calcium ATPase activity was reduced by 42% in postexercise muscle as compared to preexercise muscle.

The study showed an inverse relationship between muscle relaxation speed and Calcium ATPase. This finding supports the hypothesis that slower muscle relaxation in response to exercise-induced muscle fatigue is related to a depression in the Calcium ATPase activity.[98]

CHRONIC FATIGUE SYNDROME

Chronic fatigue syndrome is a complex disorder characterized by extreme fatigue with no definable underlying medical condition. Frustratingly, this condition does not improve with rest. It can be a devastating disease because affected individuals often function at a substantially lower level of activity than a similarly aged person without the condition. Given the importance of intracellular calcium in muscle function and fatigue, researchers measured Calcium ATPase levels in individuals suffering from chronic fatigue syndrome. More specifically, researchers were interested in whether or not Calcium ATPase levels differed between chronic fatigue patients and normal controls. Muscle biopsies were taken from both groups, and Calcium ATPase levels were 40% lower in the muscles of chronic

98 Gollnick P, Korge P, Karpakka J, Saltin B. Elongation of skeletal muscle relaxation during exercise is linked to reduced calcium uptake by the sarcoplasmic reticulum in man. Acta Physiol Scand 1991,142, 135–136.

fatigue patients as compared to the control group.[99] The authors of the study speculate that increased oxidative stress associated with chronic fatigue syndrome may result in the oxidation of Calcium ATPase and that reduced Calcium ATPase may play a role in chronic fatigue muscle function.

MUSCULAR DISEASE AND CALCIUM ATPASE

Given the importance of intracellular calcium in muscle function, it's no surprise that several devastating neuromuscular diseases are associated with disturbed calcium homeostasis. These diseases include Duchenne muscular dystrophy and ALS (amyotrophic lateral sclerosis, also known as Lou Gehrig's disease). Duchenne muscular dystrophy is a genetic disease resulting in an absence of dystrophin, a protein that helps keep muscle cells intact, while ALS is a progressive neurodegenerative disease affecting the nerves from the brain and the spinal cord that control muscles. Although the underlying causes of the diseases differ, a common endpoint for both diseases is muscle damage due to elevated intracellular calcium levels. Associated with this increased intracellular calcium level is a reduction in Calcium ATPase activity.[100,101] A ray of hope is actually on the horizon for Duchenne muscular dystrophy patients. Recent animal studies have demonstrated that increasing Calcium ATPase levels through genetic manipulation significantly reduces muscle deterioration in mouse models of Duchenne muscular dystrophy.[102]

99 Fulle S, Belia S, Vecchiet J, Morabito C, Vecchiet L, Fano G. Modification of the functional capacity of sarcoplasmic reticulum membranes in patients suffering from chronic fatigue syndrome. Neuromuscular Disorders 13 (2003) 479–84.

100 Chin E, Chen D, Bobyk K, Mazala D. Perturbations in intracellular Ca2+ handling in skeletal muscle in the G93A*SOD1 mouse model of amyotrophic lateral sclerosis. Am J Physiol Cell Physiol 2014 Dec 1; 307(11):C1031–38.

101 Schneider J, Shanmugam M, Gonzalez J, Lopez H, Gordon R, Fraidenraich D, Babu G. Increased sarcolipin expression and decreased sarco(endo) plasmic reticulum Ca2+ uptake in skeletal muscles of mouse models of Duchenne muscular dystrophy. J Muscle Res Cell Motil 2013 Dec; 34(5–6) 349–56.

102 Mazala D, Pratt S, Chen D, Molkentin J, Lovering R, Chin E. SERCA1 overexpression minimizes skeletal muscle damage in dystrophic mouse models. Am J Physiol Cell Physiol 2015 May 1; 308(9):C699–709.

EXERCISE TRAINING AND CALCIUM ATPASE

Although Calcium ATPase is reduced immediately following intense exercise, a consistent exercise training program increases Calcium ATPase over time. The training programs that researchers have shown to benefit Calcium ATPase include interval, aerobic, and weight training, and it can improve Calcium ATPase levels for people of all ages. We go into more detail both in terms of research and what it means for you in the program section.

FUTURE DIRECTIONS

- In 2019, researchers reported that treatment with a small molecule Calcium ATPase activator reversed muscle damage in an animal model of oxidative stress (which mimics age-associated muscle loss and muscular dystrophies). After seven weeks of treatment, a 23% reduction in muscle size and force was completely reversed in treated mice.[103]
- In 2018, Italian researchers reported that Electrical Muscle Stimulation (EMS) delivers beneficial effects on aging muscle. Fifteen human subjects received EMS for twenty-four sessions over nine weeks. At the end of the study, Calcium ATPase levels were upregulated in the EMS-treated group.[104]

103 Qaisar R, Bhaskaran S, Ranjit R, Sataranatarajan K, Premkumar P, Huseman K, Van Remmen H. Restoration of SERCA ATPase prevents oxidative stress-related muscle atrophy and weakness. Redox Biol. 2019 Jan;20:68-74. doi: 10.1016/j.redox.2018.09.018. Epub 2018 Sep 27. PMID: 30296699; PMCID: PMC6174848.

104 Mosole S, Zampieri S, Furlan S, Carraro U, Löefler S, Kern H, Volpe P, Nori A. Effects of Electrical Stimulation on Skeletal Muscle of Old Sedentary People. Gerontol Geriatr Med. 2018 Apr 10;4:2333721418768998. doi: 10.1177/2333721418768998. PMID: 29662923; PMCID: PMC5896842.

CHAPTER SUMMARY

1. Skeletal muscles are voluntary, as opposed to involuntary, and enable us to move.
2. Skeletal muscle contraction and relaxation are controlled by intracellular calcium levels. Thus, Calcium ATPase plays a key role in skeletal muscle function.
3. Calcium regulation and reduced Calcium ATPase also play a role in chronic fatigue syndrome and muscular dystrophy.
4. When muscles are exercised to fatigue, Calcium ATPase levels decline, resulting in short-term reduced muscle strength and slower muscle relaxation.
5. Endurance, interval, and strength training can all increase Calcium ATPase levels over time in both skeletal muscles and in the heart.

CHAPTER TWELVE

SLEEP

Ahh, sleep. The reward at the end of each day that recharges our batteries and prepares us for tomorrow. Most of us participate in this important activity each night in a regular, alternating rhythm with our waking endeavors without thinking much about the process. But let's spend the next several pages exploring sleep a bit more deeply to gain a better understanding about what actually takes place and, of course, to learn about the central role Calcium ATPase plays. Adequate Calcium ATPase is necessary for the pineal gland to efficiently produce melatonin, the key hormone that regulates sleep

THE PINEAL GLAND

The most important factor helping induce and maintain sleep states is the production of the hormone *melatonin* by our body's pineal gland, which happens in a rhythmic, daily pattern. Calcium ATPase plays an important role in melatonin production as well as the health of the pineal gland.

Melatonin induces drowsiness to help us fall asleep, then helps maintain the sleep state until the time to wake up. Here's how the process works. As light levels fall each evening, that information is relayed from our eyes to an area of the brain called the *hypothalamus,* whose primary job is to translate information from the nervous system into signals that can be recognized by the endocrine glands.

At a certain level of darkness, the hypothalamus sends a signal to the pineal gland to start producing more melatonin. The secreted melatonin activates specific parts of the brain to induce sleep and its associated bodily activities. While it is dark, the signal continues to be sent and melatonin continues to be produced, helping to keep us asleep. As light levels begin to rise in the early morning, the signaling stops and melatonin production drops.

One way to think of natural melatonin is as a sleeping pill produced by our bodies that is far superior to anything we can find at a pharmacy. Pharmaceutical sleep-drug dosing requires an educated guess as to how much is needed for an entire night's sleep and is made of chemicals that may take the liver several steps to process. When the body produces our sleep hormone internally, it has the advantage of being able to make just the right amount in tiny increments throughout the night and in a chemical form that is easy on our livers, so the quality and quantity of our sleep is well served.

CALCIUM ATPASE AND THE PINEAL GLAND

In the pineal gland cells, an increase in intracellular calcium stimulates the production of N-acetyltransferase activity (NAT). NAT is the enzyme responsible for converting serotonin into melatonin. However, this is a refined process that requires Calcium ATPase to regulate intracellular calcium levels such that the correct amount of NAT is produced. With reduced Calcium ATPase, intracellular calcium levels go awry, which ultimately leads to cellular damage that hinders NAT production and thereby melatonin. As a result, your good night's sleep goes out the window.

Let's look at some research. One study examined melatonin production in hamsters that were genetically modified to have reduced Calcium ATPase levels. The researchers measured nighttime melatonin levels and found that melatonin levels in the Calcium ATPase-modified rats were approximately 40% lower than the melatonin levels in a control group. The researchers hypothesized that the decrease in melatonin production was due to elevated intracellular calcium levels. The elevated intracellular calcium levels reduced NAT activity, which is the key factor in melatonin production.[105]

Another role Calcium ATPase plays in melatonin production is the health of the pineal gland itself. As we age, calcium deposits begin to appear in the pineal gland (called *pineal calcification*).[106] As a result of pineal calcification, pineal cells gradually degenerate, die, and break down. Subsequently, the pineal gland's ability to produce melatonin decreases. Several studies have suggested that reduced Calcium ATPase levels likely contribute to pineal gland calcification.[107,108]

MELATONIN AND CALCIUM ATPASE

THERE'S MORE TO THE STORY

In the section above, I discussed the importance of Calcium ATPase in melatonin production, but a somewhat surprising part of the story is that once produced, melatonin actually stimulates production of Calcium ATPase in other parts of the body. Three examples are in the heart, the brain, and the pancreas.

THE HEART

Researchers were interested in determining whether melatonin levels had an effect on Calcium ATPase levels in the heart. In order to isolate the effect of melatonin, they compared three groups of rats. One group consisted of rats that were given a pinealectomy (the removal of the pineal gland), the second group was kept in continuous light to eliminate melatonin production, and the third group served as a control. The study found that cardiac Calcium ATPase levels in the control group were approximately 30% higher as compared to the rats in the melatonin-starved states.[109] In short, melatonin is heart-friendly.

105 Reiter R, White T, Lerchi A, Stokkan K, Rodriguez C. Attenuated nocturnal rise in pineal and serum melatonin in a genetically cardiomyopathic Syrian hamster with a deficient calcium pump. J Pineal Res 1991: 11:156–162.
106 Mahlberg R, Kienast T, Hadel S, Heidenreich J, Schmitz S, Kunz D. Degree of pineal calcification (DOC) is associated with polysomnographic sleep measures in primary insomnia patients. Sleep Medicine 10 (2009) 439–45.
107 Krstic R. Pineal calcification: its mechanism and significance. J Neural Transm Suppl 1986; 21:415–32.
108 Krstić R. Ultracytochemical localization of calcium in the superficial pineal gland of the Mongolian gerbil. J Pineal Res. 1985;2(1):21-37. doi: 10.1111/j.1600-079x.1985.tb00625.x. PMID: 3831299.
109 Chen L, Tan D, Reiter R, Yaga K, Poeffeler B, Kumar P, Manchester L, Chambers J. In vivo and in vitro effects of the pineal gland and melatonin on Ca2+Mg2+ dependent ATPase in cardiac sarcolemma. J Pineal Res. 1993:14:178–83.

THE BRAIN

Excessive alcohol ingestion can significantly depress brain Calcium ATPase levels. Researchers were interested if pretreatment with supplemental melatonin could protect the brain from alcohol's detrimental effect on Calcium ATPase. In order to determine this, they measured Calcium ATPase levels in rat brain synapses after acute alcohol toxicity with and without pretreatment with melatonin. The result found that pretreatment with melatonin significantly prevented the acute alcohol-induced reduction in brain Calcium ATPase levels.[110]

THE PANCREAS

Acute pancreatitis (associated with alcohol abuse) is a deadly disease that has no effective treatment. Significant prior research has linked pancreatitis to excessive intracellular calcium levels in pancreatic cells. Researchers were interested if melatonin could help protect pancreatic cells from calcium overload through increased Calcium ATPase levels.

To understand this, they induced pancreatitis in two groups of rats. One group of rats was given melatonin; the other group was not given melatonin. Pancreatic Calcium ATPase levels in rats given melatonin were approximately 50% higher as compared to the rats in the group with no melatonin supplementation. The researchers concluded that melatonin appears to be capable of reducing pancreatic damage in acute pancreatitis through the stimulation of Calcium ATPase.[111]

CHRONIC OBSTRUCTIVE SLEEP APNEA

Obstructive sleep apnea (OSA) is a sleep disorder during which there are recurrent episodes of partial or complete upper airway collapse. This is dangerous because when the airways collapse, there are intermittent periods with reduced oxygen levels, called *hypoxia*. OSA is a

110 Oner P, Cinar F, Kocak H, Gurdol F. Effect of exogenous melatonin on ethanol-induced changes in Na(+), K(+) and Ca(2+)-ATPase activities in rat synaptosomes. Neurochem Res 2002 Dec; 27(12):1619–23.
111 Huai J, Shao Y, Sun X, Jin Y, Wu J, Huang Z. Melatonin ameliorates acute necrotizing pancreatitis by the regulation of cytosolic Ca2+ homeostasis. Pancreatology 12 (2012) 257–63.

major risk factor for many cardiovascular complications such as high blood pressure, stroke, atrial fibrillation, and heart failure.[112]

Although it has not been a focal point of research thus far, Calcium ATPase recently made an appearance in an OSA-related study. To study OSA under controlled conditions, researchers needed to come up with an animal model of OSA. They were able to do this by designing a tube with a special bag that could be alternatively inflated or deflated, resulting in periods of chronic intermittent hypoxia, like what occurs in human cases of sleep apnea. After three months, all rabbits in the obstruction group manifested sleep patterns similar to apnea patients. In addition, stress biomarkers were measured, and one for the purposes of this book is worth noting. After twelve weeks of engineered apnea, Calcium ATPase levels in cardiac tissue declined by approximately 44% in the treated animals as compared to the control group.[113]

112 Young, T, et al. Epidemiology of obstructive sleep apnea: a population health perspective." American journal of respiratory and critical care medicine vol. 165,9 (2002): 1217–39. doi:10.1164/rccm.2109080.
113 Xu, Li-Fang, et al. Establishment of a Rabbit Model of Chronic Obstructive Sleep Apnea and Application in Cardiovascular Consequences. Chinese medical journal vol. 130,4 (2017): 452–59. doi:10.4103/0366-6999.199828.

CHAPTER SUMMARY

1. Sleep plays an essential function during which many restorative activities take place. Among these are tissue growth and repair, immune system maintenance, rebalancing of appetite hormones, and the consolidation of memories. (There are many more!)
2. Sleep is regulated by the pineal gland through its production of a hormone called melatonin.
3. In the pineal gland, melatonin is produced by the stimulation of what is termed NAT activity.
4. When there is a reduction in Calcium ATPase, NAT activity is reduced. Therefore, melatonin is reduced and sleep quality deteriorates.
5. Reduced Calcium ATPase can also result in pineal calcification. This causes pineal cells to gradually degenerate and die, resulting in reduced melatonin production.
6. Melatonin is also a powerful antioxidant and has been shown to increase Calcium ATPase in the heart, the brain, and the pancreas.
7. Chronic sleep apnea reduces Calcium ATPase levels in the heart.

SECTION 3

The Enemies of Healthy Calcium ATPase

Here I cover the effects of environmental toxins, pesticides, and other contaminants in our food and water on Calcium ATPase. In addition, the drugs we ingest, as well as our stress levels, also have an impact on our Calcium ATPase. You will come away with a new appreciation for just how important it is to be aware of what we eat, drink, and breathe.

CHAPTER THIRTEEN

ENVIRONMENTAL TOXINS

In this chapter, I discuss thirteen common environmental toxins that reduce Calcium ATPase. I'll discuss where they are found and which Calcium ATPase disease states they are linked to. Some you will likely be familiar with, like lead and mercury. Others will be new to you.

These are the two primary ways toxins inhibit Calcium ATPase:

- Toxins create free radicals, which damage Calcium ATPase through oxidation.
- Substances such as fluoride can actually displace calcium from the enzyme, eliminating its ability to pump calcium into the sarcoplasmic reticulum.

Below is a summary of environmental toxins and chemicals that reduce Calcium ATPase in various parts of the body. From there, we will take a closer look at each toxin and where it is found.

COMPOUND	INHIBITS CALCIUM ATPASE IN . . .
Lead	brain, red blood cells, and skeletal muscles
Mercury	brain, heart, and skeletal muscles
Aluminum	brain and skeletal muscles

Fluoride	brain and kidneys
Cadmium	brain, red blood cells, arteries, liver, kidney, and testes
Chlorine	heart
Titanium Dioxide	brain, heart, skin, sperm, and skin
Zinc Oxide	retinas of the eyes, and in the skin
Flame retardants	brain, skeletal muscle, and sperm
Trichlorethylene	heart
Carbon Tetrachloride	liver, brain, and red blood cells
Polychlorinated Bisphenols	brain

LEAD

Studies have shown that lead specifically inhibits Calcium ATPase levels in the brain, red blood cells, and skeletal muscles.[114,115,116]

According to the EPA, streams utilized by water utility companies that provide water to a third of the US population are not yet covered by clean-water laws that limit levels of toxic pollutants such as lead. That's over one hundred million people who could potentially be drinking from lead-tainted water. Moreover, even if the water supply is clean, lead pipes, when damaged, can leach lead into tap water.

Unsafe levels of lead exposure in children are associated with significant developmental delays, brain damage, learning and behavioral problems (including hyperactivity), kidney damage, and even death. Even at very low lead levels, the deleterious effects of reduced Calcium ATPase can mean children may have problems

114 Bettaiya R, Yallapragada PR, Hall E, Rajanna S. In vitro effect of lead on Ca(2+)-ATPase in synaptic plasma membranes and microsomes of rat cerebral cortex and cerebellum. Ecotoxicol Environ Saf. 1996 Mar; 33(2):157–62.
115 Sandhir R, Gill KD. Alterations in calcium homeostasis on lead exposure in rat synaptosomes. Mol Cell Biochem. 1994 Feb 9; 131(1):25–33.
116 Hechtenberg S, Beyersmann D. Inhibition of sarcoplasmic reticulum Ca(2+)-ATPase activity by cadmium, lead and mercury. Enzyme. 1991; 45(3):109–15.

with attention, learning, and controlling impulses. The Academy of Pediatricians recommends that because most US children are at risk of lead exposure, their blood levels should be tested at least once during childhood.

The negative effects that lead has on Calcium ATPase have been known for decades. In 1994, researchers studied how lead disrupts calcium regulation in brain synapses. Rats were exposed to lead for eight weeks. As expected, levels of lead in the blood increased as compared to the control group. In addition, Calcium ATPase levels in the brain synapses were reduced by approximately 30% in the treated rats, and intracellular calcium levels were 90% higher than the control group. Researchers concluded that the calcium disruption caused by lead could lead to such problems as nervous system dysfunction (including memory and learning) as well as cell death.[117]

A 2008 study focused on the effect that lead had on Calcium ATPase levels in workers who were exposed regularly. The study group consisted of thirty male employees who had been exposed to lead at least ten years. The control group was twenty individuals who had not been exposed to lead on a regular basis. As expected, the lead count was much higher in the exposed workers. In fact, it was approximately thirty times higher than the control group (317 ug/dl compared to 10 ug/dl). In terms of Calcium ATPase, the exposed group had significantly lower Calcium ATPase levels, while the control group had approximately four times the amount of Calcium ATPase as compared to the exposed group. The authors conclude that the detrimental effect lead has on Calcium ATPase leads to toxicity.[118]

Another important study investigated the effects of low-level environmental lead exposure on Calcium ATPase. The researchers measured lead levels and Calcium ATPase in 247 mothers and newborns by testing strands of their hair. The results were startling. Even though the women's blood levels were under the "toxic" levels (90% below the "admissible limit" in most countries), their Calcium ATPase levels, and perhaps more profoundly the newborns' Calcium ATPase levels, were correlated with their lead exposure.

117 Sandhir R, Gill KD. Alterations in calcium homeostasis on lead exposure in rat synaptosomes. Mol Cell Biochem. 1994;131(1):25–33. doi:10.1007/BF01075721.
118 Yücebilgiç G, Bilgin R, Tamer L, Tükel S. Effects of lead on Na(+)-K(+) ATPase and Ca(+2) ATPase activities and lipid peroxidation in blood of workers. Int J Toxicol. 2003;22(2):95–97.

The more lead in the mother's hair, the lower the Calcium ATPase. The researchers concluded that even levels of lead below the acceptable limit may result in delayed neuropsychological development of newborns and children.[119]

Sources of Exposure

In addition to contaminated drinking water, lead exposure can occur via the following:

- **Paint:** Although lead paint was banned for indoor use in 1978, according to the American Academy of Pediatrics, 25% of children still live in households with deteriorated lead-based paint. The dust and chips from this paint expose these children to a proportionally higher level of lead. Paint used at daycare centers as well as schools can also contribute. Many homes built before 1978 contain lead-based paint.
- **Soil:** Lead can be found in soil from exterior paint or a past exposure to leaded gas—another good reason for taking off shoes before coming inside the home.
- **Inexpensive jewelry:** From 2004 to 2007, manufacturers recalled more than forty-five jewelry products involving 170 million units due to excessive lead. In 2006, a four-year-old Minneapolis boy died after swallowing a trinket he received as a free gift with his new Reebok sneakers—the trinket contained more than 90% lead. The incident brought to light the fact that many American toy companies had been violating federal safety standards for almost thirty years, and new regulation was put in place.
- **Toys manufactured before 2008:** Before 2008, lead content in children's toys was not tightly regulated. In response to massive toy recalls in 2007—including 1.5 million Thomas and Friends wooden railway toys and nearly a million Fisher Price toys—the United States Consumer Product Safety Improvement Act (CPSIA) was

119 Campagna D, Huel G, Hellier G, et al. Negative relationships between erythrocyte Ca-pump activity and lead levels in mothers and newborns. *Life Sci.* 2000;68(2):203–15.

signed into law in 2008. The CPSIA specifically targets any product designed or intended primarily for use by children twelve years of age or younger. Testing and certification of children's products to the requirements of the CPSIA are required for legal entry into the US market.

- **Vintage toys, cribs, and furniture:** Many vintage toys from the 1970s and '80s, such as dollhouses, blocks, wagons, Mattel Barbie dolls, and many Fischer Price brand toys, were painted with lead paint. However, you may be surprised to learn that plastic toys from this era also contained lead and cadmium. A study analyzed lead and cadmium content from this time and found 67% of vintage toys exceeded current American and European limits. The good news: Tests on contemporary toys did not contain measurable levels of lead or cadmium.
- **Hobbies:** Making pottery or stained glass, or mechanical projects that involve soldering, may expose you to lead.

MERCURY

Studies have shown that excess levels of mercury are associated with reduced Calcium ATPase in the brain, skeletal muscles, and heart.[120,121,122]

You've most likely heard of the negative effects caused by mercury exposure. The avoidance of seafood with high mercury levels, the near elimination of mercury fever thermometers for home use, and the reduced use of mercury in dental fillings are the most recent manifestations of this new public awareness. Exposure to this toxin is especially devastating for infants during the prenatal period and in newborns during breastfeeding. In particular, the brain and nervous system of infants may be damaged permanently, resulting in the impairment of memory, attention, language, fine

120 Freitas AJ, Rocha JB, Wolosker H, Souza DO. Effects of Hg2+ and CH3Hg+ on Ca2+ fluxes in rat brain microsomes. Brain Res. 1996 Nov 4; 738(2):257–64.

121 Hechtenberg S, Beyersmann D. Inhibition of sarcoplasmic reticulum Ca(2+)-ATPase activity by cadmium, lead and mercury. Enzyme. 1991; 45(3):109–15.

122 Furieri LB, Fioresi M, Junior RF, Bartolomé MV, Fernandes AA, Cachofeiro V, Lahera V, Salaices M, Stefanon I, Vassallo DV. Exposure to low mercury concentration in vivo impairs myocardial contractile function. Toxicol Appl Pharmacol. 2011 Sep 1; 255(2):193–99.

motor skills, and visual special skills. Let's look at the Calcium ATPase connection.

In 1996, in an effort to understand the mechanism whereby mercury causes damage, researchers investigated the effect of mercury on calcium regulation in the brain, specifically in the cerebral cortex and the cerebellum. To this end, rat brain synapses and microsomes were exposed to micromolar concentrations of mercury compounds, and Calcium ATPase activity was then measured. The experiment found that mercury and methylmercury both significantly reduced Calcium ATPase levels in a concentration-dependent manner.[123] A different study in 1996 found that mercury had a negative impact on Calcium ATPase levels in rat brain microsomes.[124]

A more recent 2008 study investigated the effect that mercury levels in mothers had on their newborns' Calcium ATPase levels. A total of ninety-eight mothers participated in the study, in which mercury levels of the mother's hair were compared to the Calcium ATPase levels in the newborn. The results indicated that small changes in maternal mercury levels had a significant impact on Calcium ATPase levels in newborns.

The takeaway: even incremental changes in mercury level can have a negative impact on Calcium ATPase, which is an important enzyme that plays a significant role neuronal development.[125] It makes sense why the EPA has recommendations discouraging expectant mothers from eating fish, such as swordfish, that contain high mercury levels.

The damage mercury can cause doesn't stop there. A 2011 study investigated the effect that chronic low-level mercury exposure had on the heart. The study focused on several markers of cardiac function, including its effect on Calcium ATPase. Rats were treated for thirty days with mercury to mimic chronic environmental exposure (blood levels 8ml as compared to EPA safety range of 5.8 ml). After thirty days, several of the cardiac markers were unchanged, i.e., no change in blood pressure, or enlargement of the heart. However,

123 Yallapragada P, Rajanna S, Fail S, Rajanna B. Inhibition of Calcium Transport by Mercury Salts in Rat Cerebellum and Cerebral Cortex In Vitro. Journal of Applied Toxicology, Vol 16(4),325–30 (1996).
124 Freitas AJ, Rocha JB, Wolosker H, Souza DO. Effects of Hg2+ and CH3Hg+ on Ca2+ fluxes in rat brain microsomes. Brain Res. 1996;738(2):257–264. doi:10.1016/s0006-8993(96)00781-0.
125 Huel G, Sahuquillo J, Debotte G, Oury JF, Takser L. Hair mercury negatively correlates with calcium pump activity in human term newborns and their mothers at delivery. Environ Health Perspect. 2008;116(2):263–267.

there was an approximate 20% reduction in Calcium ATPase levels. The researchers concluded that chronic exposure to mercury, even at low concentrations, is an environmental risk factor affecting heart function.[126]

Sources of Exposure

There are two basic forms of mercury, both of which occur naturally. *Elemental mercury* is found in the Earth's crust and in our atmosphere. Approximately 50% of mercury in the atmosphere comes from the natural emissions of volcanic eruptions. Humans contribute the other 50%, primarily from coal-fired power plants. Exposure to elemental mercury can occur through dental amalgam fillings (which can be up to 90% mercury), broken mercury-filled thermometers, broken fluorescent lightbulbs, and some equipment in school laboratories. We also breathe in mercury from air polluted by the aforementioned coal-powered plants.

When mercury in the atmosphere falls to the Earth in rain, it then accumulates in lakes and rivers and is converted to *methylmercury*, which is elemental mercury converted by microorganisms in the sea. Methylmercury is then absorbed into the bodies of fish and seafood. Through the process of biomagnification, as bigger fish eat littler fish, the bigger fish end up with higher levels of methylmercury. Those bigger fish then end up on our dinner plates. Thus, our greatest exposure to methylmercury comes from the seafood we consume. For a list that outlines the safest and most dangerous varieties of fish and seafood, please see the program chapters.

126 Furieri L, Fioresi M, Ribeiro R, Bartolome M, Fernandes A, Cachofeiro V, Lahera V, Salaices, Stefanon I, Vassallo D. Exposure to low mercury concentration in vivo impairs myocardial contractile function. Toxicology and Applied Pharmacology 255 (2011) 193–99.

ALUMINUM

Studies show that aluminum has a negative effect on Calcium ATPase in the brain and skeletal muscles.[127,128,129,130]

Aluminum is the most abundant metal in the Earth's crust and is found naturally in air, water, and soil. It's also used in a wide variety of products with which we come into contact every day. You can't completely eliminate aluminum from your environment, as it's too ubiquitous. In fact, the average person consumes and eats over three pounds of aluminum in their lifetime—the equivalent of 229 square feet of aluminum foil! Research has clearly demonstrated that prenatal and newborn exposure to excessive aluminum results in neurodevelopmental defects and significant long-term effects on bone health. In adults, elevated aluminum concentrations in the brain and nervous system have been reported in several neurodegenerative and muscle-wasting diseases, including Alzheimer's, Parkinson's, and amyotrophic lateral sclerosis (ALS). Let's look at the Calcium ATPase connection.

In a 1994 study, primates were exposed to aluminum through intragastric dosing every other day for fifty-two weeks. At the end of the fifty-two weeks, Calcium ATPase levels were measured in the cerebral cortex, the hippocampus, and the corpus stratum. In all three areas of the brain, Calcium ATPase levels were significantly reduced as compared to the control group. The biggest decline occurred in the cerebral cortex, with 30% reduction of Calcium ATPase. As would be expected, the reduced Calcium ATPase resulted in increased intracellular calcium levels that were 30% higher in all parts of the brain. The researchers concluded that the elevated calcium levels could have adverse effects on synaptic transmission, dendrite induction, and axonal transport leading to obvious severe neuronal damage.[131]

---

127 Kaur A, Gill KD. Disruption of neuronal calcium homeostasis after chronic aluminium toxicity in rats. Basic Clin Pharmacol Toxicol. 2005 Feb; 96(2):118–22.

128 Sarin S, Julka D, Gill KD. Regional alterations in calcium homeostasis in the primate brain following chronic aluminium exposure. Mol Cell Biochem. 1997Mar;168(1-2):95–100.

129 Julka D, Gill KD. Altered calcium homeostasis: a possible mechanisms of aluminium-induced neurotoxicity. Biochim Biophys Acta. 1996 Jan 17;1315(1):47–54.

130 de Sautu M, Saffioti NA, Ferreira-Gomes MS, Rossi RC, Rossi JPFC, Mangialavori IC. Aluminum inhibits the plasma membrane and sarcoplasmic reticulum Ca(2+)-ATPases by different mechanisms. Biochim Biophys Acta Biomembr. 2018 Aug;1860(8):1580–88.

131 Sarin S, Julka D, Gill KD. Regional alterations in calcium homeostasis in the primate brain following chronic aluminium exposure. *Mol Cell Biochem*. 1997;168(1-2):95–100.

A 1998 study confirmed these findings in rat neurons with Calcium ATPase levels significantly reduced by aluminum exposure.[132] Further evidence is provided by a 2004 study that investigated how twelve weeks of aluminum exposure would affect the brains of the exposed rats. The study found that Calcium ATPase levels were reduced up to 25% in certain parts of the brain. As expected, intracellular calcium levels were elevated over 100% in certain parts of the brain. The researchers concluded that the reduced Calcium ATPase activity and the resulting elevation in intracellular calcium could be the underlying mechanisms that trigger aluminum toxicity, which may play a role in neurodegenerative diseases such as Alzheimer's, Parkinson's, and ALS.[133]

Sources of Exposure

In addition to the obvious sources of aluminum, such as cookware, cans, and aluminum-based deodorant, you may be surprised to know that many baked goods contain aluminum-based baking powder, and certain antacids are aluminum based. Aluminum foil, if used to cook acidic ingredients like tomatoes, citrus, and BBQ sauce, can seep aluminum into your food. See the program chapters for more specific ways to reduce aluminum exposure.

FLUORIDE

Studies have shown that fluoride has a negative effect on Calcium ATPase levels in the brain, skeletal muscle, and kidneys.[134,135,136]

Similar to lead and aluminum, fluoride occurs naturally in the Earth's crust, water, and air, making it a ubiquitous component of our environment.

132 Gandolfi L, Stella MP, Zambenedetti P, Zatta P. Aluminum alters intracellular calcium homeostasis in vitro. Biochim Biophys Acta. 1998;1406(3):315–20.
133 Kaur A, Gill K. Disruptioin of Neuronal Calcium Homeostasis after Chronic Aluminum Toxicity in Rats. Basic and Clinical Pharmacology and Toxicology 2005,96,118–22.
134 Zhao XL, Gao WH, Zhao ZL. Effects of sodium fluoride on the activity of Ca2+Mg(2+)-ATPase in synaptic membrane in rat brain. Zhonghua Yu Fang Yi Xue Za Zhi. 1994 Sep; 28(5):264–66.
135 Murphy AJ, Coll RJ. Fluoride is a slow, tight-binding inhibitor of the calcium ATPase of sarcoplasmic reticulum. J Biol Chem. 1992 Mar 15; 267(8):5229–35.
136 Borke JL, Whitford GM. Chronic fluoride ingestion decreases 45Ca uptake by rat kidney membranes. J Nutr. 1999 Jun; 129(6):1209–13.

It's safe to assume that you are familiar with brushing your teeth regularly with fluoride toothpaste to protect your teeth from cavities. Fluoride does protect teeth enamel and serves a valuable function, especially during the period when permanent teeth are developing. Fluoride used externally without ingestion is low risk. However, the chronic ingestion of fluoride is a completely different story. It has been linked to a wide range of health problems, including bone disease, kidney dysfunction, reduced cognition, and lower IQ in children.

In fact, in large doses, fluoride is a poison. Need proof? Just look at your toothpaste tube. You'll see a poison warning clearly displayed, courtesy of an FDA requirement. In its purest form, 1/10 of an ounce of fluoride can kill a one-hundred-pound adult; 1/100 of an ounce can kill a ten-pound baby. Let's look closer at the connection between fluoride and Calcium ATPase.

In a 1994 study, researchers investigated the effects of fluoride on Calcium ATPase activity in the rat brain. There were two components: an "in vitro" test, which involved exposing isolated rat brain cells to fluoride in a test tube, and an "in vivo" test, which involved giving fluorinated water to female rats during gestation and lactation for fifty days. The results were that fluoride can significantly inhibit the activity of Calcium ATPase both in the test tube and direct animal studies. The test tube study found Calcium ATPase reductions of up to 63% at the highest level of exposure. In the animal test, Calcium ATPase activity was reduced 11%–32% in both the mother and their newborns as compared to the control group.[137]

In 1999, researchers investigated the effects of fluoride on Calcium ATPase activity in kidney cells. Female rats were given fluorinated water at different dosages for six weeks. Exposure levels were selected that would produce plasma concentrations of fluoride attainable by humans from environmental exposure. In all treatment groups, Calcium ATPase activity was reduced as compared to the control group. The study suggests that chronic high fluoride ingestion may result in calcium dysregulation due to its negative effect on Calcium ATPase.[138]

---

137 Zhao XL, Gao WH, Zhao ZL. Zhonghua Yu Fang Yi Xue Za Zhi. 1994;28(5):264–66.
138 Borke JL, Whitford GM. Chronic fluoride ingestion decreases 45Ca uptake by rat kidney membranes. J Nutr. 1999;129(6):1209–13.

Sources of Exposure

Fluoride is a good thing when it helps to keep the enamel on our teeth strong and protective. However, it does not make sense to add fluoride to our drinking water. How could it possibly make sense to intentionally ingest a Calcium ATPase inhibitor multiple times a day? In fact, dental fluorosis (unattractive discoloration of the teeth) occurs during the first eight years of a child's life when permanent teeth are being formed. Its cause is too much exposure to fluoride and is seen in about 65% of adolescents in the United States.

Another reason to avoid drinking fluoride: as I detailed in the chapter on obesity, Calcium ATPase plays a significant role in metabolism. In short, Calcium ATPase burns calories via the mechanism of uncoupled pumping without actually transporting calcium. This is called uncoupled transport. As it turns out, fluoride impairs this uncoupled Calcium ATPase activity. In other words, fluoride lowers your metabolism.

Along with dental products and water, fluoride is also found in food and drink that use fluoridated water for processing. For example, one serving of chicken sticks contains 50% of the upper limit for safe levels of fluoride for a one-year-old toddler.

Some farmers use pesticides that contain fluoride to kill produce-destroying insects. Fluoride residue is then found in related foods, such as cocoa powder, walnuts, and grape products. A good reason to buy organic.

CADMIUM

Studies have shown that increased exposure to cadmium has adverse effects on Calcium ATPase, specifically in the brain, red blood cells, arteries, skeletal muscle, liver, intestines, and testes.[139,140,141,142,143,144,145,146]

Cadmium is a basic element found naturally in the Earth. If you don't remember it from high school chemistry classes, you will have seen it on the label of your nickel-cadmium batteries and many other common consumer products. Exposure to cadmium, a toxic heavy metal, has been shown to be linked to kidney disease, hypertension, atherosclerosis, developmental problems in children (including low IQ), and cancer. All of these disease states exist in areas where cadmium has a negative impact on Calcium ATPase. Let's look closer at the link between cadmium and Calcium ATPase.

A study in 1987 looked at the effect cadmium had on calcium regulation in the intestine. They did this by measuring the impact of cadmium on rat intestinal epithelial cells. They discovered something quite compelling—cadmium has a high affinity for the same binding sites as calcium on Calcium ATPase. In other words, cadmium and calcium are competing for the same spot. As a result, Calcium ATPase can't do its job effectively and intracellular calcium levels skyrocket, leading to a host of cellular problems.[147]

In 1991, researchers investigated the effect cadmium had on Calcium ATPase activity in the brain by exposing isolated rat brain cells to cadmium. Exposed cells had 15%–60% lower levels of Calcium ATPase activity than control cells. The study's data clearly

---

139 Vig PJ, Nath R. In vivo effects of cadmium on calmodulin and calmodulin regulated enzymes in rat brain. Biochem Int. 1991 Mar; 23(5):927–34.

140 Vig PJ, Nath R, Desaiah D. Metal inhibition of calmodulin activity in monkey brain. J Appl Toxicol. 1989 Oct; 9(5):313–16.

141 Verbost PM, Flik G, Pang PK, Lock RA, Wendelaar Bonga SE. Cadmium inhibition of the erythrocyte Ca2+ pump. A molecular interpretation. J Biol Chem. 1989 Apr 5; 264(10):5613-15.

142 Sumida M, Hamada M, Takenaka H, Hirata Y, Nishigauchi K, Okuda H. Ca2+,Mg2+-ATPase of microsomal membranes from bovine aortic smooth muscle: effects of Sr2+ and Cd2+ on Ca2+ uptake and formation of the phosphorylated intermediate of the Ca2+,Mg2+-ATPase. J Biochem. 1986 Sep; 100(3):765–72.

143 Hechtenberg S, Beyersmann D. Inhibition of sarcoplasmic reticulum Ca(2+)-ATPase activity by cadmium, lead and mercury. Enzyme. 1991; 45(3):109–15.

144 Zhang GH, Yamaguchi M, Kimura S, Higham S, Kraus-Friedmann N. Effects of heavy metal on rat liver microsomal Ca2(+)-ATPase and Ca2+ sequestering. Relation to SH groups. J Biol Chem. 1990 Feb 5; 265(4):2184–89.

145 Verbost PM, Senden MH, van Os CH. Nanomolar concentrations of Cd2+ inhibit Ca2+ transport systems in plasma membranes and intracellular Ca2+ stores in intestinal epithelium. Biochim Biophys Acta. 1987 Aug 20; 902(2):247–52.

146 Murugavel P, Pari L. Diallyl tetrasulfide modulates the cadmium-induced impairment of membrane bound enzymes in rats. J Basic Clin Physiol Pharmacol. 2007; 18(1):37–48.

147 Verbost P, Senden M, van Os CH. Nanomolar concentrations of Cd2+ inhibit Ca2+transport systems in plasma membranes and intracellular Ca2+ stores in intestinal epithelium. Biochemica et Biophysica Acta 902 (1987) 247–52.

demonstrate that cadmium acts as a Calcium ATPase inhibitor.[148] A 2007 study took it a step further and looked at how cadmium would affect the brain of rats dosed with cadmium over a period of three weeks. At the end of the three weeks, Calcium ATPase levels were 30% lower in the treated rats as compared to the control group.[149]

Sources of Exposure

Exposure to cadmium can come from a number of sources. One of great concern is jewelry. Now that lead is banned because of its toxic effects, cadmium is used to add weight and extra shine to those tiny bobbles and bangles. This is especially problematic when used in children's jewelry.

In 2010, Walmart was notified by AP reporters that a line of Miley Cyrus jewelry contained potentially dangerous levels of cadmium. Unfazed, the retailer continued to sell the line along with other jewelry that had been reported to contain high levels of cadmium. The problem isn't the risk of cadmium leaching into the skin, it's the probability of kids chewing on and accidentally swallowing the jewelry, causing them to ingest unsafe levels of cadmium. In testing conducted for the AP, some Chinese-made children's jewelry contained as much as 91% cadmium. These items were (and possibly still are) sold in America . . . to children . . . at Walmart. That same year in the United States, the McDonald's restaurant chain voluntarily recalled twelve million Shrek-themed children's drinking glasses due to their high content of the toxic metal. Since then, regulations in the use of cadmium have been tightened somewhat in children's toys and jewelry (though more laws need to be passed).

There are no laws regulating cadmium in adult jewelry, and cadmium-containing products are currently sold at national retailers including Ross, Nordstrom Rack, and Papaya. In 2018, analysis done for the nonprofit Center for Environmental Health found that some jewelry sold with women's dresses and shirts was nearly pure cadmium. Overall, the lab results found thirty-one adult jewelry items purchased from retail stores that were at least 40% cadmium; most were more than 90%.

---

148 Shah J, Pant H. Effect of cadmium on Ca2+ transport in brain microsomes. Brain Research 566 (1991)127–30.

149 Pari L, Murugavel P. Diallyl tetrasulfide improves cadmium induced alterations of acetylcholinesterase, ATPases and oxidative stress in brain of rats. Toxicology 234 (2007) 44–50.

Cadmium exposure occurs from other sources, as well. Traces of cadmium are found in tobacco plants, thus making cigarettes a source of cadmium exposure. Most people who smoke have about twice as much cadmium in their blood as nonsmokers. Secondhand smoke can also expose passersby to cadmium. In nonsmokers, the primary source of cadmium is food. Cadmium is released into the air from various industrial sources and upon dissipation settles into soil and oceans. Fish and plants, and the animals that eat those plants, absorb cadmium. The most common cadmium-exposed food sources are cereals, nuts, vegetable oils, and meats.

CHLORINE

Studies have shown that chlorine vapors negatively affect Calcium ATPase levels in the heart.

Chlorine is yet another naturally occurring element of our Earth. For many years, its use as a disinfectant has protected us from bacteria and viruses in water, potentially saving millions of lives. As a cleaner, it disinfects and sanitizes areas where people work, eat, and live. However, studies have shown that while chlorine protects us, this common substance has also been harming us, sometimes lethally.

In gas form, chlorine can cause birth defects, miscarriages, immune disorders, and, according to some sources, cancer. Chlorinated drinking water, while saving us from bacteria and viruses, can produce carcinogenic trihalomethanes.

Let's look at the Calcium ATPase connection. Other than respiratory complications, more than 90% of fatalities due to high-level chlorine exposure may result from cardiovascular damage. Researchers were interested in why. Given the importance of Calcium ATPase in cardiac function, a 2015 study investigated the impact exposure to chlorine gas had on Calcium ATPase levels in the heart. Rats were exposed for thirty minutes to a high concentration of chlorine gas. After exposure, Calcium ATPase levels in heart cells were measured. As compared to the control group, Calcium ATPase levels declined by over 50%. Results show that the negative effect chlorine gas has on Calcium ATPase levels in the heart is a major component of chlorine-related cardiac damage.[150]

Sources of Exposure

We absorb chlorine into our bodies one of three ways: through our skin, by ingesting it, or by breathing in the fumes. Those long showers you love to take, letting the entire room fog up beautifully? If you have chlorinated, unfiltered water, you're breathing in chlorine gas, a potentially deadly substance. In fact, chlorine has been labeled a chemical weapon by the International Community of the Red Cross and was used as a deadly weapon in World War I by Germany, France, and Great Britain. Today, many sanitation departments refuse to dispose of full containers of chlorine bleach because it's considered hazardous material.

Yes, this is the same stuff you probably have in your cleaning cabinet at home. When combined with dish soap, the two will react to create mustard gas. In addition, when mixed with other chemicals found in bathroom cleaning products, carbon tetrachloride—a Calcium ATPase inhibitor—can form. We will cover this inhibitor in a later section.

Part of the problem is that because of its cleansing benefits, chlorine is extremely common. It's in our swimming pools, cleaning products, teeth whiteners, and hair colorings.

Especially troubling is the promotion of bleach handy wipes to wipe down classrooms. The chlorine smell or vapor often promoted as the smell of a clean germ-free environment is actually toxic. Furthermore, in order to actually kill bacteria and viruses effectively, bleach must saturate a *precleaned surface* (bleach itself is not an effective cleaner) for five to ten minutes.

150 Ahmad S, Ahmad A, Hendry-Hofer TB, et al. Sarcoendoplasmic reticulum Ca(2+) ATPase. A critical target in chlorine inhalation-induced cardiotoxicity. Am J Respir Cell Mol Biol. 2015;52(4):492–502.

TITANIUM DIOXIDE

Titanium dioxide nanoparticles have been shown to adversely affect Calcium ATPase levels in the brain, heart, sperm, and epithelial cells when exposed to UV rays.

You might remember lifeguards at the beach or pool sitting in the sun all day long with whitened noses. That was titanium dioxide in its large particle form, which has never been a concern. Today, however, we use titanium dioxide made of much smaller particles. So small, in fact, that they have been reduced to nanoparticle size. To put this in perspective, one nanoparticle ranges from 1 to 100 nm. A human hair ranges from 80,000 to 100,000 nms wide. So, with a steady hand, you could stack about 1,000 nanoparticles on the end of a strand of hair. This helps explains why it is more easily absorbed than previous versions. Nanoparticle titanium dioxide is used in sunscreen, and to add whiteness to paints, cosmetics, and food.

These tiny particles can easily enter our bodies, travel into our bloodstream, inhibit Calcium ATPase, and cause harm. They are in so many products that you've probably eaten titanium dioxide nanoparticles and/or worn them on your skin within the last week, if not the last day. Public health and environmental advocates have sounded the alarm over the widespread use and safety of titanium dioxide nanoparticles, particularly since published studies have indicated that titanium dioxide nanoparticles are toxic to the brain, heart, lung, liver, ovary, testes, spleen, and kidney of animals. Wow, what a list! Let's look at its connection to Calcium ATPase.

A 2010 study set out to determine the neurotoxicological effects and impairment of spatial recognition memory caused by exposure to titanium dioxide nanoparticles. Mice were exposed to the nanoparticles through intragastric administration every day for sixty days. (Keep in mind that we humans ingest titanium dioxide nanoparticles regularly through numerous processed foods.) At the end of sixty days, significant neurological effects were seen, including a profound reduction in performance on a memory test. The researchers reported a significant decrease in Calcium ATPase levels, which contributed to the negative neurological effects of the titanium dioxide nanoparticle exposure.[151]

A second 2010 study looked at the interaction between titanium dioxide nanoparticles and ultraviolet light (UV) in lens epithelial cells. Researchers reported that upon exposure to UV rays, titanium dioxide nanoparticles created free radicals, reduced Calcium ATPase, and caused the elevation of intracellular calcium leading to the disruption of intracellular calcium homeostasis, which induced cell damage and death.[152]

In a 2015 study, researchers reported a connection between titanium dioxide nanoparticle exposure and biological dysfunction during spermatogenesis. In the study, male mice were exposed to titanium dioxide nanoparticles through intragastric administration for sixty consecutive days. Exposed mice developed lesions of the testis, reductions in sperm concentration and motility, and an increase in the number of abnormal sperm in mice. After measuring a number of biomarkers, researchers reported up to a 57% decrease in Calcium ATPase levels. This reduction in Calcium ATPase played a role in sperm function.[153]

A 2016 study found a connection between titanium dioxide nanoparticles and Calcium ATPase in the heart. Mice were continuously exposed to a low dose of titanium dioxide nanoparticles for ninety days via intragastric administration. At the end of the ninety days, a range of biomarkers were tested, including Calcium ATPase. Calcium ATPase levels was significantly reduced, which contributed to the negative effects of titanium dioxide particles on the heart.[154]

---

151 Hu R, Gong X, Duan Y, Li N, Che Y, Cui Y, Zhou M, Liu C, Wang H, Hong F. Neurotoxicological effects and the impairment of spatial recognition memory in mice caused by exposure to TiO2 nanoparticles. Biomaterials 2010 Nov;31(31):8043–50.

152 Wu Q, Guo D, Du Y, Liu D, Wang D, Bi H. UVB irradiation enhances TiO2 nanoparticle-induced disruption of calcium homestasis in human lens epithelial cells. Photochem Photobiol 2014 Nov-Dec;90(6):1324–31.

153 Hong F, Si W, Zhao X, Wang L, Zhou Y, Chen M, Ge Y, Zhang Q, Wang Y, Zhang J. TiO2 nanoparticle exposure decreases spermatogenesis via biological dysfunctions in the testis of male mice. J Agric Food Che 2015 Aug 12;63(31):7084–92.

154 Yu X, Hong F, Zhang YQ. Cardiac inflammation involving in PKCe or ERK1/2-activated NF-kB signaling pathway in mice following exposure to titanium dioxide nanoparticles. J Hazard Mater. 2016 Aug 5;313:68–77.

Sources of Exposure

Part of the problem is that titanium dioxide exposure doesn't end with just putting on suntan lotion or eating a donut. Because of its excessive use, it enters our sewage systems and works its way into our soil and water supplies. We then ingest more titanium dioxide nanoparticles through our meat, fish, and vegetables.

Unfortunately, children are particularly susceptible to titanium dioxide nanoparticle exposure because of its use in candy and chewing gum. There are many, many food items and medications that list titanium dioxide as an ingredient. Although most do not designate nanoparticle-sized titanium dioxide, researchers have found that approximately 35% of titanium dioxide found in food is nanoparticle in size.

Below, we also discuss the dangers of titanium dioxide nanoparticles, zinc oxide nanoparticles, and other chemicals in sunscreen.

ZINC OXIDE

Studies have shown that zinc oxide nanoparticles have a negative effect on Calcium ATPase in the retinas of the eyes and in epithelial cells when exposed to UV light.

You are likely aware of the use of zinc oxide in sunscreen products. Just like titanium dioxide, zinc oxide sunscreen in the old days had a trademark thick, white consistency. These days, zinc oxide nanoparticles make zinc oxide sunscreens nearly invisible.

The problem with zinc oxide nanoparticles is the same as that with titanium dioxide: they are small enough to be absorbed through the skin and into the circulatory system, where they can then reach the liver, kidney, heart, and spleen. Animal studies have found that zinc oxide nanoparticles induce liver damage and result in a significant rise in inflammatory markers and oxidative stress. In addition, other studies have found that zinc oxide nanoparticles are toxic to cells in the colon. Let's look at the Calcium ATPase connection.

A 2013 study looked at the combined effect of zinc oxide nanoparticles and exposure to ultraviolet light in human lens epithelial cells. The results indicated that the combination of zinc oxide nanoparticles and ultraviolet light resulted in elevated intracellular calcium, disruption in calcium homeostasis, and reduction in the plasma membrane Calcium ATPase levels.[155]

A second 2013 study was interested in the mechanism by which zinc oxide nanoparticles become toxic to various organisms. The researchers exposed rat retinal ganglion cells (vital neurons that convey information from other retinal neurons to the rest of the brain) to zinc oxide nanoparticles. Researchers found that zinc oxide nanoparticles caused the creation of free radicals, a reduction in plasma membrane Calcium ATPase, and a disruption in intracellular calcium, which triggers mitochondrial dysfunction and cell death.[156]

Sources of Exposure

There is danger, particularly with children, when applying sunscreen with zinc oxide nanoparticles on the lips, mouth area, and around the nose, and also via food that has been handled by sunscreened hands. Another danger lies in spray sunscreen and the inhalation of titanium dioxide and zinc oxide nanoparticles into the sensitive lining of the lungs.

Test results from a 2018 research study commissioned by Friends of the Earth found engineered nanoparticles in all four major US children's sunscreen products tested. These products included: Aveeno Baby Natural Protection (Nano titanium and Nano zinc oxide); Banana Boat Kids (Nano titanium and Nano zinc oxide); Neutrogena Pure and Fresh Baby (Nano zinc oxide); and Thinksport Kid's Safe (Nano zinc oxide). We will give you some non-nano sunscreen options for kids in the program chapters.

More on sunscreens

Most squamous cell cancers appear in people over fifty, but in recent years, more cases are being diagnosed in people in their twenties and thirties. In particular, the number of women under age forty diagnosed with squamous cell cancer has significantly increased in

155 Wang D, Guo D, Bi H, Wu Q, Tian Q, Du Y. Zinc oxide nanoparticles inhibit Calcium ATPase expression in human lens epithelial cells under UVB irradiation. Toxicol in Vitro 2013 Dec;27(8):2117–26.

156 Guo D, Bi H, Wang D, Wu Q. Zinc oxide nanoparticles decrease the expression and activity of plasma membrane calcium ATPase, disrupt the intracellular calcium homeostasis in the rat retinal ganglion cells. Int J Biochem Cell Biol 2013 Aug;45(8):1849–59.

the last thirty years. This is a puzzling statistic given the increased popularity and availability of daily lotions, make-ups, and other beauty products that contain sunscreen. While the answer is likely complex (including, potentially, the use of tanning beds), it could be that the ingredients in sunscreens are a contributing factor.

Three commonly used ingredients—titanium dioxide, zinc oxide, and oxybenzone—have demonstrated clear effectiveness in blocking ultraviolet light. However, they have also been shown to disrupt Calcium ATPase under photooxidizing conditions (via ultraviolet light). This is troubling because research shows that skin cancer is associated with reduced Calcium ATPase levels.

Titanium dioxide nanoparticles have been shown to reduce Calcium ATPase in skin cells, while zinc oxide nanoparticles have been shown to inhibit Calcium ATPase under ultraviolet light in skin cells and the retina. Oxybenzone (benzophenone 3) has been shown to significantly reduce Calcium ATPase activity from 15% to 30% under ultraviolet light. When Calcium ATPase is exposed to light without benzophenone, there is no reduction in Calcium ATPase activity.

Ultimately, all three of these ingredients that are meant to protect us from skin cancer could actually be increasing our likelihood of getting skin cancer. The CDC has detected oxybenzone in 96% of Americans. Given the importance of Calcium ATPase in skin cancer, further research needs to be done to ensure that this unintended effect of key sunscreen ingredients is not contributing to the development of the disease they are meant to prevent.

FLAME RETARDANTS

Studies have shown that brominated fire retardants negatively affect Calcium ATPase in the brain, skeletal muscle, and sperm.[157,158,159] Flame retardant compounds come in two common varieties, halogenated and PBDE. They both accumulate in our bodies through dust exposure and inhalation, causing damage to reproductive and

157 Al-Mousa F, Michelangeli F. The sarcoplasmic-endoplasmic reticulum Ca(2+)-ATPase (SERCA) is the likely molecular target for the acute toxicity of the brominated flame retardant hexabromocyclododecane (HBCD). Chem Biol Interact. 2014 Jan 25; 207:1–6.

158 Ogunbayo OA, Michelangeli F. The widely utilized brominated flame retardant tetrabromobisphenol A (TBBPA) is a potent inhibitor of the SERCA Ca2+ pump. Biochem J. 2007 Dec 15; 408(3):407–15.

159 Ogunbayo OA, Lai PF, Connolly TJ, Michelangeli F. Tetrabromobisphenol A (TBBPA), induces cell death in TM4 Sertoli cells by modulating Ca2+ transport proteins and causing dysregulation of Ca2+ homeostasis. Toxicol In Vitro. 2008 Jun; 22(4):943–52.

neurological systems, among others. In an unfortunate twist, new fire safety regulation passed in the 1970s has ensured that these very same fire retardants are more common than ever.

In 2014, a group of researchers at the Environmental Working Group partnered with scientists at Duke University to test twenty-two mothers and twenty-six children. Fire retardants were found in the urine of every mother and child tested. In the children, the average concentration of a fire-retardant metabolite was approximately five times the average in the mothers. The lesson: fire retardant exposure is real, and children are the most affected.

In 2007, researchers conducted the first study that examined the relationship between flame retardants and Calcium ATPase. The study focused on one particular compound called tetrabromobisphenol A (TBBPA), which is currently the most widely used brominated fire retardant. They measured the effect TBBPA had on Calcium ATPase activity in skeletal muscles and brain neurons and determined that TBBPA was a potent inhibitor of Calcium ATPase in both cell types.[160]

In 2008, researchers were interested in TBBPA's effect on sperm development. In order to do this, they exposed mouse Sertoli cells (essential for sperm development). They found that exposure of Sertoli cells to TBBPA significantly decreased Calcium ATPase and in concert led to a significant increase in intracellular calcium, mitochondrial disturbance, and cell death.

In 2011, researchers at the University of Birmingham in the United Kingdom conducted a study to see if brominated flame retardants could be a risk factor in neurodegenerative disease, such as Alzheimer's. They examined what happened to brain neurons when exposed to three fire retardants: TBBPA, hexabromocyclododecane (HBCD), and decabromodiphenyl ether (DBPE). The study found that all three chemicals were cytotoxic at extremely low concentrations. Neurons that were exposed to these compounds had a 25%–80% reduction in Calcium ATPase, with correspondingly elevated intracellular calcium levels. The elevated intracellular calcium levels appear to trigger mitochondrial dysfunction and ultimately cell death. Elevated intracellular calcium levels may also contribute

160 Ogunbayo OA, Michelangeli F. The widely utilized brominated flame retardant tetrabromobisphenol A (TBBPA) is a potent inhibitor of the SERCA Ca2+ pump. *Biochem J.* 2007;408(3):407–15. doi:10.1042/BJ20070843.

to the increase in Amyloid beta peptide, a key pathological marker in Alzheimer's.[161]

These results were replicated in a 2013 study in which researchers examined a total of seven flame-retardant compounds. All seven tested were neurotoxic. Furthermore, the researchers found a direct correlation between their potency in inducing cell death and their ability to inhibit Calcium ATPase. They conclude that the acute toxicity of these compounds comes from the dysregulation of intracellular calcium as a result of reduced Calcium ATPase activity.[162]

Sources of Exposure

Fire retardants are more common than we first assume. Because they are invisible to the naked eye, it may be hard to imagine that exposure is really something we should be concerned about.

For example, flame retardants are added to polyurethane foam blocks founds in gym foam pits. When the foam is compressed by landing on it, the flame retardants escape into the air and attach to dust particles. These dust particles then attach themselves to athletes' or children's hands and bodies. A recent study conducted at Duke University found that hand-wipe samples showed the individuals exposed to the foam pits had higher levels of the chemicals on their skin after practice than before practice. Gymnasts in particular had four to six times more flame retardants in the blood than in the US general population.

At home, flame retardants lurk in older furniture cushions, carpet padding, and foam mattress pads. The flame retardant leaks from these items and sticks to dust. The contaminated dust collects on household surfaces, such as toys and food, and is then inhaled or ingested. Kids are the most likely to be exposed because of their tendency to put toys and their hands into their mouths.

Think of how often a newborn baby is swaddled in a soft comfy blanket that meets fire safety standards and therefore contains toxic fire retardants. Those dust bunnies under your couch and bed?

161 Al-Mousa F, Michelangeli F. Some commonly used brominated flame retardants cause Ca2+-ATPase inhibition, beta-amyloid peptide release and apoptosis in SH-SY5Y neuronal cells. *PLoS One*. 2012;7(4):e33059. doi:10.1371/journal.pone.0033059.

162 Al-Mousa F, Michelangeli F. The sarcoplasmic-endoplasmic reticulum Ca(2+)-ATPase (SERCA) is the likely molecular target for the acute toxicity of the brominated flame retardant hexabromocyclododecane (HBCD). Chem Biol Interact. 2014;207:1–6.

Filled with fire retardant chemicals that any exploring child or pet will come into contact with. Your child's crib mattress probably has the same chemicals.

TRICHLOROETHYLENE

Studies have shown that trichloroethylene negatively impacts Calcium ATPase in the heart, particularly during the prenatal period.[163,164,165] Trichloroethylene is a colorless, volatile liquid used in the production of refrigerants, fire retardants, pharmaceuticals, and insecticides. It is also used as a solvent to remove grease from metal parts and as a spot cleaner in dry cleaning. The EPA classifies trichloroethylene as carcinogenic to humans by all routes of exposure. It is toxic to the kidneys, liver, immune system, male reproductive system, and the developing fetus (potentially contributing to fetal heart malformations). Let's take a look at the Calcium ATPase connection.

Trichloroethylene (TCE) exposure is linked to congenital heart malformations in both animal and human studies. Heart defects are the number one cause of infant death in the United States, accounting for 1/5 of total infant deaths. In 1994, researchers investigated the connection between TCE and cardiac defects in newborns. Pregnant rats were exposed to various levels of TCE in drinking water from ten to twelve days. Embryo DNA was collected, and tests were done to search for gene deviations associated with TCE exposure. The gene coding for Calcium ATPase in rat embryonic cells was one of two genes that were downregulated consistently in embryos exposed to TCE. In fact, Calcium ATPase gene levels declined 18–90% over the dosage levels tested. In other words, Calcium ATPase reduction was a significant indicator of TCE exposure and could explain the association of TCE with neonatal heart defects given the importance of Calcium ATPase in cardiac development.[166]

---

163 Hoffmann P, Heinroth K, Richards D, Plews P, Toraason M. Depression of calcium dynamics in cardiac myocytes—a common mechanism of halogenated hydrocarbon anesthetics and solvents. J Mol Cell Cardiol. 1994 May; 26(5):579–89.

164 Caldwell PT, Thorne PA, Johnson PD, Boitano S, Runyan RB, Selmin O. Trichloroethylene disrupts cardiac gene expression and calcium homeostasis in rat myocytes. Toxicol Sci. 2008 Jul; 104(1):135–43.

165 Palbykin B, Borg J, Caldwell PT, Rowles J, Papoutsis AJ, Romagnolo DF, SelminOI. Trichloroethylene induces methylation of the Serca2 promoter in H9c2 cells and embryonic heart. Cardiovasc Toxicol. 2011 Sep; 11(3):204–14.

166 Collier JM, Selmin O, Johnson PD, Runyan RB. Trichloroethylene effects on gene expression during cardiac

In 2008, researchers at the University of Arizona conducted a study to determine if calcium dysregulation by TCE could play a role in the negative cardiac effects of TCE exposure. Rat cardiac cells were exposed to TCE for twelve hours or twenty-four hours. After twelve hours, Calcium ATPase levels declined 22%, and after forty-eight hours, levels declined by 43%.[167]

A 2011 study investigated how exposure to low doses of TCE affects physiological heart development in the embryonic heart. The study reported that low doses of TCE exposure had a negative effect on Calcium ATPase (through a process called DNA methylation) in embryonic heart cells. The negative effect of TCE on Calcium ATPase could be a key mechanism that explains the cardiac defects associate with TCE.[168]

Sources of Exposure

TCA is also found in consumer products such as metal degreasers, arts and crafts sprays, sealants, adhesives, paints, paint strippers, varnishes, pepper sprays, and hoof polishes. Trichloroethylene may be present in drinking water, indoor environments, surface water, groundwater, and soil. In 2015 alone, 172 facilities disposed of approximately two million pounds of trichloroethylene into the environment.[169]

CARBON TETRACHLORIDE

Carbon tetrachloride was a popular aerosol propellant from the 1950s to1970s and appeared in dry-cleaning agents, refrigerants, cleaners, and pesticides. Due to its toxicity and negative greenhouse gas effects, carbon tetrachloride's use and production has been greatly curtailed over the last twenty years—from approximately one billion pounds in 1974 to approximately 142 million pounds in 2015. Let's look at the Calcium ATPase connection.

---

development. Birth Defects Res A Clin Mol Teratol. 2003;67(7):488–95.

167 Caldwell PT, Thorne PA, Johnson PD, Boitano S, Runyan RB, Selmin O. Trichloroethylene disrupts cardiac gene expression and calcium homeostasis in rat myocytes. *Toxicol Sci.* 2008;104(1):135–43.

168 Palbykin B, Borg J, Caldwell PT, et al. Trichloroethylene induces methylation of the Serca2 promoter in H9c2 cells and embryonic heart. *Cardiovasc Toxicol.* 2011;11(3):204–14.

169 https://saferchemicals.org/get-the-facts/toxic-chemicals/tce-trichloroethylene/.

The negative effects of carbon tetrachloride on the liver are well known, and research has shown a connection to Calcium ATPase. A 1991 study found that rats treated with carbon tetrachloride had a 50% reduction in red blood cell Calcium ATPase levels.[170] The results suggested that the Calcium ATPase levels in red blood cells could be a simple, safe, and useful marker of early liver damage. A 1998 study investigated the impact of a single oral dose of carbon tetrachloride on Calcium ATPase levels in the liver. The rats treated with carbon tetrachloride had a significant decrease in liver Calcium ATPase activity and a significant increase in intracellular calcium as compared to the control group.[171] A similar rat study was conducted in 2008, and Calcium ATPase activities in red blood cells showed a significant decline in treated rats as compared to the control group.[172]

Sources of Exposure

Carbon tetrachloride is still used in some products, including industrial adhesives and tapes, paints used for things like swimming pools and traffic lines, arts and crafts paste, carpet spot removers, paint strippers, paint removers, and brake cleaners. It can also be emitted in vapor form from the use of cleaning products that contain bleach. Remember not to mix chlorine bleach with other cleaning products, because carbon tetrachloride can be formed.

POLYCHLORINATED BISPHENOLS (PCBS)

Polychlorinated bisphenols (PCBs) are a group of man-made chemicals used in the production of paints, adhesives, flame retardants, plasticizers, calking compounds, lubricating fluids, heat transformers, capacitators, and pesticides. In other words, a lot of stuff! PCBs are highly toxic. Babies and children are particularly vulnerable because the metabolic pathway necessary to eliminate PCBs from their bodies has not fully developed, and exposure to PCBs may

170 Mourelle M, Franco M. Erythrocyte defects precede the onset of CCI$-induced liver cirrhosis. Protection by silymarin. Life Sci 1991; 48(11): 1083–90.
171 Katsumata T, Murata T, Yamaguchi M. Alteration in calcium content and Calcium ATPase actvitiy in the liver nuclei of rats orally administered carbon tetrachloride. Mol Cell Biochem 1998 Aug; 185(1-2): 153–59.
172 Akyuz F, Aydin O, Ali Demir T, Kanback G. The effects of CCI4 on Na+/K+ATPase and trace elements in rats. Biol Trace Elem Res 2009 Dec;13291-3):207–14.

result in developmental and neurological problems. Let's look at the connection between PCBs and Calcium ATPase.

In a 2007 study, researchers reported the effects of PCBs on two parts of the brain, the hypothalamus and the hippocampus. After thirty days of exposure, measurements of various biomarkers, including Calcium ATPase, were compared to those of a control group. Calcium ATPase levels were significantly decreased. Some good news, though: researchers also had a group that received Vitamin E supplementation along with the PCBs, and the combination treatment resulted in Calcium ATPase levels that were close to normal. We will discuss this further in the upcoming program chapters.

A 2008 study also confirmed the negative effect a thirty-day PCB exposure had on Calcium ATPase levels in several parts of the brain, including the cerebellum, cerebral cortex, and hippocampus. Again, good news: one group of rats was given melatonin in addition to PCBs, which resulted in the prevention of PCB reduction in Calcium ATPase.

Sources of Exposure

In 1977, PCB production was banned. However, more than 1.5 billion pounds of PCBs were manufactured in the US before 1977. PCBs are slow to break down and remain an environmental toxin decades later. So slow, in fact, that PCB is termed a persistent bioaccumulative toxin. Due to the chemical structure of PCBs, they tend to settle at the bottom of lakes and oceans. Bottom-feeding fish accumulate PCBs from sediment, and the PCBs become more concentrated as they move up the food chain. As a result, PCB levels in some fish can be as much as one million times higher than PCB levels in the water. Fish and shellfish are our primary sources of exposure to this toxin.

CHAPTER FOURTEEN

PESTICIDES

The Centers for Disease Control and Prevention reports that there are traces of twenty-nine different pesticides in the average American's body.[173] Below is a chart that summarizes various pesticides and their impact on Calcium ATPase. That may be as far as you need to read. In the program section, I will give you a plan to reduce pesticide exposure in your food and home. However, if you are interested, I have more in-depth information on the long-term health effects of specific pesticides.

Regardless of how far you delve into each section, the takeaway is clear: pesticides have a negative impact on Calcium ATPase.

PESTICIDE	INHIBITS CALCIUM ATPASE IN
Chlorpyrifos	brain, skeletal muscle, red blood cells
Dichlorvos	brain, lungs
Parathion	heart, brain, central nervous system, red blood cells
Dimethoate	skeletal muscle, red blood cells
Cypermethrin	brain

173 https://www.consumerreports.org/cro/health/natural-health/pesticides/index.htm.

Deltatmethrin	brain
Permethrin	brain and red blood cells
Resmethrin	brain
Atrazine	heart, liver, red blood cells
Paraquat	liver
Benthiocarb	brain
DDT	brain, liver, red blood cells, ovary, placenta
Toxaphene	brain, heart, kidney, liver, red blood cells
Chlordecone	brain, skeletal muscle, heart
Endosulfan	brain, red blood cell, sperm
Dieldrin	brain, heart
Lindane	skeletal muscle, testes

ORGANOPHOSPHATE PESTICIDES
CHLORPYRIFOS, DICHLORVOS, DIMETHOATE, AND PARATHION

Studies demonstrate that organophosphate pesticides have a negative impact on Calcium ATPase in the brain, skeletal muscle, red blood cells, and lungs.[174,175,176,177,178,179,180]

The most compelling danger of organophosphate pesticides overall is their effect on children. One study of eight-to-fifteen-year-olds found that those with the highest urinary levels of a marker for exposure to organophosphates (OPs) had twice the odds of developing attention deficit hyperactivity disorder as those with undetectable levels.[181] Another study found that at age seven, children of California farmworkers born to mothers with the highest levels of OPs in their bodies while pregnant had an average IQ that was seven points below those whose mothers had the lowest levels during pregnancy. That's comparable to the IQ loss children suffer due to low-level lead exposure.[182]

Chlorpyrifos

Leading environmental safety groups have battled for years to ban chlorpyrifos, an organophosphate pesticide used to kill termites, fire ants, mosquitoes, and various other insects that damage agricultural crops. A study by Columbia University found that ". . . even low to moderate levels of exposure to the insecticide chlorpyrifos during pregnancy may lead to long-term, potentially irreversible changes in the brain structure of the child."[183]

174 Choudhary S, Gill K. Protective effect of nimodipine on dichlorvos-induced delayed neurotoxicity in rat brain. Biochem Pharmacol 2001 Nov, 1:62(9):1265–72.

175 Raheja G, Gill K. Calcium homeostasis and dichlorvos induced neurotoxcity in rat brain. Mol Cell Biochem 2002 Mar; 232)1–2):13–18.

176 Cul J, Li C, He X, Song Y. Protective effects of penehyclidine hydrochloride on acute lung injury caused by severe dichlorvos poisoning in swine. Chin Med J (Engl)2013; 126 (24):4764–70.

177 Barber D, Hunt J, Ehrich M. Inhibition of calcium-stimulated ATPase in the hen brain P2 synaptosomal fraction by organophosphorus esters: relevance to delayed neuropathy. J. Toxicol Environ Health A, 2001 May 25; 63(2);101–13.

178 Mehta A, Verma R, Srivastaya N. Chlorpyrifos-induced alterations in rat brain acetylcholinesterase, lipid peroxidation and ATPases. Indian J Biochem Biophys, 2005 Feb; 42(1):54–58.

179 Singh M, Sandhir R, Kiran R. Erytrhocyte antioxidant enzymes in toxicological evaluation of commonly used organophosphate pesticides. Indian J Expo Biol 2006 Jul; 44(7):580–83.

180 Nozdrenko D, Korchinska L, Soroca V. Activity of Ca2+,Mg(2+)-ATPase of sarcoplasmic reticulum and contraction of strength of the frog skeletal muscles under the effect of organophosphorus insecticides. Ukr Biochem J 2015 Jul-Aug: 87(4):83–89.

181 Yu CJ, Du JC, Chiou HC, Chung MY, Yang W, Chen YS, Fuh MR, Chien LC, Hwang B, Chen ML. Increased risk of attention-deficit/hyperactivity disorder associated with exposure to organophosphate pesticide in Taiwanese children. Andrology. 2016 Jul; 4(4):695–705.

182 Bouchard MF, Chevrier J, Harley KG., Kogut K, Vedar, M, Calderon N, Trujillo C, Johnson C, Bradman A, Barr DB, and Eskenazi B (2011). Prenatal Exposure to Organophosphate Pesticides and IQ in 7-Yxear-Old Children. Environmental Health Perspectives, 119(8), 1189–95.

183 https://www.mailman.columbia.edu/public-health-now/news/prenatal-exposure-insecticide-chlorpyrifos-linked-alterations-brain-structure.

According to the EPA, the largest agricultural market for chlorpyrifos is corn, but it is also used on soybeans, fruit and nut trees, brussels sprouts, cranberries, broccoli, and cauliflower, as well as other row crops. Nonagricultural uses include golf courses, turf, greenhouses, and nonstructural wood treatments such as utility poles and fence posts. It is also registered for use as a mosquito adulticide, and for use in roach and ant bait stations.[184]

Dichlorvos

The Agency for Toxic Substances and Disease Registry, or ATSDR, which is a group within the US Department of Health and Human Services, has issued a warning against a far too commonly used pesticide, dichlorvos. Dichlorvos adversely affects the nervous system in both adults and children. To make matters worse, several agencies, such as the EPA and IARC, have stated dichlorvos is a possible or probable carcinogen.[185,186] Those insect strips you might have used for bed bugs, silverfish, cockroaches, or spiders? Check the label, as they might have dichlorvos as the active ingredient.

Dimethoate

Dimethoate has been named as a potential carcinogen, and it is known to cause organ damage after chronic exposure. It is used to kill mites and insects on a wide variety of fruits, vegetables, and grains, such as wheat, apples, grapes, tomatoes, and many more. It is also used as a residual wall spray in farm buildings to kill house flies and is administered to livestock for control of botflies.[187]

---

184 https://www.epa.gov/ingredients-used-pesticide-products/chlorpyrifos.
185 https://www.atsdr.cdc.gov/toxfaqs/tfacts88.pdf.
186 Rastogi SK, Tripathi S, Ravishanker D. A study of neurologic symptoms on exposure to organophosphate pesticides in the children of agricultural workers. Indian J Occup Environ Med. 2010 Aug; 14(2):54–57. doi: 10.4103/0019-5278.72242.
187 http://pmep.cce.cornell.edu/profiles/extoxnet/dienochlor-glyphosate/dimethoate-ext.html.

Parathion

This little gem is highly toxic. In fact, the EPA has labeled parathion as "extremely toxic," and the World Health Organization has classified it as "extremely hazardous." Primarily, overexposure to parathion affects the respiratory and nervous systems, blood, eyes, and skin. Fortunately, parathion has been banned in many countries throughout the world, although its use is still allowed in the US via aerial equipment—mostly because it is too dangerous to be handled by hand, although aerial application ignores the dangers associated with its inhalation and ingestion. Dues to its toxicity, it can be used on specific crops only, such as alfalfa, barley, corn, soybeans, and wheat.[188,189,190,191]

PYRETHROIDS

CYPERMETHRIN, DELTAMETHRIN, PERMETHRIN, RESMETHRIN

Studies demonstrate that pyrethroids have a negative impact on Calcium ATPase in the brain and red blood cells.[192,193,194,195]

Pyrethroids are a synthetic version of a natural pesticide derived from the extract of chrysanthemum flowers (pyrethrin). Pyrethroids were developed as a less-toxic alternative to organophosphate and organochlorine pesticides. However, messing with Mother Nature has again proven to be a problem. In fact, for over thirty years, a substantial body of research has linked pyrethroid exposure to neurodevelopmental/learning disorders.[196,197,198]

---

188 https://www3.epa.gov/pesticides/chem_search/reg_actions/reregistration/fs_PC-057501_1-Sep-00.pdf.
189 https://www.epa.gov/sites/production/files/2016-09/documents/parathion.pdf.
190 Blasiak J Inhibition of erythrocyte membrane (Ca2+ + Mg2+)-ATPase by the organophosphorus insecticides parathion and methylparathion. Comp Biochem Physiol C Pharmacol Toxicol Endocrinol. 1995 Feb; 110(2):119–25.
191 http://www.fao.org/docrep/W5715E/w5715e05.htm.
192 Sahib IK, Prasada RKS, Desaiah. Pyrethroid inhibition of basal and calmodulin stimulated Calcium ATPase and adenylate cyclase in rat brain. J Appl Toxicol 1987 Apr; 7(2):75–80.
193 Kodavanti P, Mundy W, Tilson H, Harry G. Effects of selected neuroactive chemicals on calcium transporting systems in rat cerebellum and on survival of cerebellar granule cells. Fundam Appl Toxicol 1993 Oct; 21(3): 308–16.
194 Deifalla H. Properties of Ca2++Mg-ATPase from rat brain and its inhibition by Pyrethroids. Pesticide Biochem and Physio 1990 37, 116–20.
195 Grosman N, Diel F. Influence of pyrethroids and piperonyl butoxide on the Ca (2+)-ATPase activity of rat brain synaptosomes and leukocyte membranes.. Int Immunopharmacol 2005 Feb; 592) 263–70.
196 https://www.ncbi.nlm.nih.gov/pubmed/29648420.
197 https://www.ncbi.nlm.nih.gov/pubmed/28811173.
198 https://www.ncbi.nlm.nih.gov/pubmed/28580452.

HERBICIDES

ATRAZINE AND PARAQUAT

Studies have shown that atrazine has a negative impact on heart, liver, and red blood cell Calcium ATPase.[199,200] Studies have also shown that paraquat has a negative impact on liver Calcium ATPase.[201]

Atrazine

The second-most common herbicide used in America today is atrazine. Atrazine is used to control weeds in corn, sorghum, sugarcane, pineapple, Christmas trees, and other crops. In fact, 75% of all corn, 58% of all sorghum, and 76% of all sugarcane crops are treated with atrazine.[202] Evidence is mounting concerning the dangers of this widely used herbicide's toxicity to humans. Epidemiological studies have linked atrazine exposure to an increased risk of miscarriage, reduced male fertility, low birth rate, elevated risk of breast and prostate cancer, and increased chance of birth defects. For obvious reasons, the European Union banned atrazine in 2004.[203,204,205,206,207]

Runoff from crops, lawns, and golf courses ends up in our drinking water, which means we are probably drinking it every single day. In fact, the USDA Pesticide Data Program found 94% of water they tested had traces of atrazine.[208] You may even be using it in your own yard, as some Scott Turf-Builder and Spectracide products contain atrazine.

---

199 Singh M, Sandhir R, Kiran R. Alterations in Ca2+ homeostasis in rat erythrocytes with atrazine treatment: positive modulation by vitamin E. Mol Cell Biochem 2010 Jul; 340(1-2):231–38.
200 Lin J, Li H, Qin L, Du Z, Xia J, Li J. A novel mechanism underlies atrazine toxicity in quails (Coturnix): triggering ionic disorder via disruption of ATPases. Oncotarget 2016 Dec 20;7(51).
201 Tadashi U, Kei-ichi H. Effects of paraquat on the mitochondrial structure and Calcium ATPase activity in rat hepatocytes. 1985 J Elect Micros; 34(2);85–91.
202 https://www.beyondpesticides.org/assets/media/documents/pesticides/factsheets/Atrazine.pdf.
203 https://www.regulations.gov/document?D=EPA-HQ-OPP-2013-0266-1160.
204 https://articles.mercola.com/sites/articles/archive/2016/10/19/atrazine-health-effects.aspx.
205 Winchester PD, Huskins J, Ying J. Agrichemicals in Surface Water and Birth Defects in the United States. Acta Paediatrica (Oslo, Norway: 1992) 98.4 (2009): 664–669. PMC. Web. 13 Sept. 2018.
206 Swan SH, et al. Semen Quality in Relation to Biomarkers of Pesticide Exposure. Environmental Health Perspectives 111.12 (2003): 1478–84. Print.
207 Winchester PD, Huskins J, Ying J. Agrichemicals in Surface Water and Birth Defects in the United States. Acta Paediatrica (Oslo, Norway: 1992) 98.4 (2009): 664–69. PMC. Web. 13 Sept. 2018.
208 http://www.panna.org/resources/atrazine.

Paraquat

An herbicide used for weeds and grass, paraquat is widely used throughout the world. In the United States, the use of paraquat has increased fourfold in the last ten years. According to the EPA, paraquat is registered for use on a plethora of vegetables, grains, and fruits. It is also used around noncrop areas to control weeds, in places like public airports, electric transformer stations, and commercial buildings.[209] In 2015 alone, seven million pounds of the herbicide were sprayed on close to fifteen million US acres. Effects from long-term exposure can include damage to the liver, kidneys, heart, and respiratory system. A 2011 National Institutes of Health (NIH) study found people who used paraquat were 2.5 times more likely to develop Parkinson's than nonusers.[210,211,212]

Paraquat is banned in thirty-two countries, including the European Union. In 2012, the government of China said that it would phase out paraquat "to safeguard people's lives."[213]

ORGANOCHLORINE PESTICIDES

DDT, TOXAPHENE, CHLORDECONE, ENDOSULFAN, DIELDRIN, AND LINDANE

Studies have shown that organochlorine pesticides have a negative impact on Calcium ATPase levels in the brain, red blood cells, developing fetuses, heart, and sperm.[214,215,216,217]

These chemicals have been considered so harmful to humans, they've been mostly banned in Europe and the United States, although you can still find them in China, some Latin American countries, Africa, and India. So if you're buying produce or their derivatives from those countries, you could also be buying and ingesting organochlorine pesticides.[218,219] More locally in the United States, they still

209 https://archive.epa.gov/pesticides/reregistration/web/pdf/0262fact.pdf.
210 https://www.nih.gov/news-events/news-releases/nih-study-finds-two-pesticides-associated-parkinsons-disease.
211 https://www.epa.gov/pesticide-worker-safety/paraquat-dichloride-one-sip-can-kill#deaths.
212 https://www.reuters.com/article/brazil-pesticide-paraquat-idUSL2N0WY2V720150402.
213 https://www.nytimes.com/2016/12/20/business/paraquat-weed-killer-pesticide.html.
214 https://www.ncbi.nlm.nih.gov/pubmed/1701552.
215 Jinna RR, Uzodinma JE, Desaiah D. Age-related changes in rat brain ATPases during treatment with chlordecone. J Toxicol Environ Health. 1989; 27(2):199–208. PubMed PMID: 2471839.
216 https://www.ncbi.nlm.nih.gov/pubmed/2974087.
217 https://www.ncbi.nlm.nih.gov/pubmed/2950239
218 https://www.ncbi.nlm.nih.gov/pmc/articles/PMC5464684/.
219 https://www.thebalancesmb.com/what-are-organochlorine-pesticides-2538275.

remain in our soil and our water supplies and remain a health hazard. While effective in killing crop-damaging bugs, they are also known for their "high toxicity, slow degradation and bio accumulation."[220] Also, one of these chemicals is still used regularly in the US in prescription drugs. In 2014, 20,000 prescriptions for lindane were filled for the treatments of scabies and lice. This, despite the fact the WHO's agency on cancer research gave lindane Group 1 status, which is "carcinogenic to humans."

220 https://www.ncbi.nlm.nih.gov/pmc/articles/PMC5464684/.

CHAPTER FIFTEEN

FOOD AND TOXINS

In this chapter, we focus on food toxins and their impact on Calcium ATPase. Two main areas are covered. First, certain food additives and preservatives have a negative impact on Calcium ATPase, including BHA/BHT, TBHQ, bisphenol and nonylphenol, potassium bromate, and food dyes. Secondly, the creation of benzo(a)pyrenes during cooking has a negative effect on Calcium ATPase

FOOD PRESERVATIVES

On the one hand, food preservatives serve a noble purpose. For example, they prevent oils in foods from oxidizing and becoming rancid. Natural substances such as Vitamin E, Vitamin A, Vitamin C, and extracts of rosemary and thyme can be used as preservatives. However, on the other hand, in humankind's quest for longer shelf life, synthetic antioxidants were developed and are now ubiquitous. BHT, BHA, and TBHQ are the three primary synthetic antioxidants currently in use. Their effect on Calcium ATPase is widespread throughout the body.

BHT, BHA, and TBHQ

Studies have shown that these preservatives have a negative effect on Calcium ATPase throughout the body, including the brain, skeletal muscles, mast cells, arteries, neurons, and more.[221,222,223,224,225,226]

In animal studies, these preservatives have been proven to have negative health effects. Long-term exposure to high doses of BHA/BHT is toxic in mice and rats, promoting the development of tumors. In fact, the NIH's National Toxicology Program reported that BHA can be "reasonably anticipated to be a human carcinogen," causing liver, thyroid, and kidney problem and affecting lung function and blood coagulation.

Furthermore, recent studies provide evidence that high doses of BHT may mimic estrogen and prevent expression of male sex hormones, resulting in adverse reproductive effects. TBHQ, which forms when BHA is metabolized in the body, is also a tumor promoter and has been shown to cause liver enlargement and neurotoxic effects in laboratory animals.

FOOD DYES

Studies have shown that Red #3 has a negative effect on Calcium ATPase in the brain, skeletal muscles, and red blood cells.[227,228,229] In addition, Red #2, Red #40, Yellow #5, Yellow #6 and Blue #1 (all food dyes currently approved for use by the FDA) inhibit mitochondrial respiration,[230] which is a key step in the production of ATP, the

221 Akasaka R, Teshima R, Ikebuchi H, Sawada J. Effects of three different Calcium ATPase inhibitors on Ca2+ response and leukotriene release in RBL-2H3 cells. Inflamm Res 1996 Dec; 45(12):583–89.
222 Akasaka R, Teshima R, Kitajima S, Momma J, Inoue T, Korokawa Y, Ikebuchi H, Sawada J. Effects of hydroquinone-type and phenolic antioxidants on calcium signals and degranulation of RBL-2H3 cells. Biochem Pharmacol 1996 Jun 14; 51(11):1513–19.
223 Teshima R, Onose J, Ikebuchi H, Sawada J. Calcium ATPase inhibitors and PKC activation synergistically stimulate TNF-alpha production in RBL-2H3 cells. Inflamm Res 1998 Aug; 47(8):328–33.
224 Teshima R, Onose J, Okunuki H, Sawada J. Effect of Calcium ATPase inhibitors on MCP-1 release from bone marrow-derived mast cells and the involvement of p38 MAP kinase activation. Int Arch Allergy Immunol 2000 Jan; 121(1):34–43.
225 Nakamura R, Ishida S, Ozawa S, Saito Y, Okunuki H, Teshima R, Sawada J. Gene expression profiling of Calcium ATPase inhibitor DTBHQ and antigen-stimulated RBL-2H3 mast cells. Inflamm Res 2002 Dec; 51(12):611–18.
226 Kitajima S, Momma J, Tsuda M, Kurokawa Y, Teshima R, Sawada J. Effects of 2,5-(tert-butyl)-1,4-hydroquinine on intracellular free Ca2+ levels and histamine secretion in RBL-2H3 cells. Inflamm Res 1995 Aug; 44(8):335–39.
227 Heffron JJ, O'Callaghan AM, Duggan PF. Food dye, erythrosin B, inhibits ATP-dependent calcium ion transport by brain microsomes. Biochem Int. 1984 Nov; 9(5):557–62.
228 Mugica H, Rega AF, Garrahan PJ. The inhibition of the calcium-dependent ATPase from human red cells by erythrosin B. Acta Physiol Pharmacol Latinoam. 1984; 34(2):163–73.
229 Morris SJ, Silbergeld EK, Brown RR, Haynes DH. Erythrosin B (USFD&C RED 3) inhibits calcium transport and atpase activity of muscle sarcoplasmic reticulum. Biochem Biophys Res Commun. 1982 Feb 26; 104(4):1306–11
230 Reyes FG, Valim MF, Vercesi AE. Effect of organic synthetic food colours on mitochondrial respiration. Food Addit Contam. 1996 Jan;13(1):5–11.

energy that our bodies run on. This is relevant to Calcium ATPase because the energy that fuels Calcium ATPase is ATP, and mitochondrial respiration is responsible for its production.

There is a whole different discussion we could have about food dyes and behavior, cancer, and birth defects. Research is ongoing and conflicting. So for this book, I will stay focused on dyes and Calcium ATPase. All you need to know is that food dyes are bad for Calcium ATPase, and that's reason enough to cut them out.

The story of Red Dye #3 is a great example of the battle between science and regulatory agencies. By the late 1970s, concerns about Red Dye #3's impact on children's neurological state, specifically its relation to hyperactivity, began to surface. Two separate studies in 1979 reported that Red Dye #3 impacted the function of the neurotransmitter dopamine in brain tissue and thus could act as a central excitatory agent able to induce hyperkinetic behavior. A 1980 study reported that Red Dye #3 produced an irreversible, dose-dependent increase in the release of the neurotransmitter acetylcholine, which could also cause excitability.

In 1982, researchers at the NIH published an article finding that "low concentrations" of Red Dye #3 significantly inhibited skeletal muscle Calcium ATPase. It suggested that exposure to Red Dye #3, and its negative impact on muscle function, might contribute to altered behavior. A study published in 1984 reported that Red Dye #3 inhibited Calcium ATPase in brain microsomes. A second study published in 1984 reported that Red Dye #3 inhibited Calcium ATPase in human erythrocytes (red blood cells).

A turning point came in 1985, when a study reported that Red Dye #3 was a thyroid tumor promoter. Then a second study followed in 1990, which also reported that chronic ingestion of Red Dye #3 promoted thyroid tumor formation. In 1990, the FDA banned the use of Red Dye #3 in cosmetics but under food industry pressure did not ban it as a food additive. At the time, there was no perfect substitute for Red Dye #3, prompting fears of dull pink maraschino cherries! In the early 1990s, the industry began shifting toward the use of Red Dye #40.

Although today Red Dye #40 (which has also been linked to hyperactivity) is the predominant red food coloring, Red Dye #3 is still being used in some foods, and the negative effects continue to pile up. In 1997, a study reported that Red Dye #3 increased sperm abnormalities and germ cell mutagenicity in rats. A 2002 study reported that Red Dye #3, along with seven other dyes, induced dose-related DNA damage in the stomach, colon, and bladder. A study in 2009 reported that a *single administration* of Red Dye #3 could inhibit the neurotransmitter serotonin in three parts of the brain.

A second study in 2009 reported that Red Dye #3 inhibited drug-metabolizing enzymes in humans. Two studies published in 2010 found that long-term Red Dye #3 exposure induced motor hyperactivity in rats. An additional 2010 study reported that Red Dye #3 causes DNA damage to human lymphocytes. A 2012 study found that Red Dye #3 caused DNA damage in human liver cells in vitro. This damage was comparable to the damage caused by a chemotherapy drug whose purpose is to break down liver DNA. A 2015 study found that in rats, prenatal exposure to commonly used food dyes, including Red Dye #3, resulted in decreased motivation and increased anxiety in offspring.

In short, any time a product that you buy is red, check the label for Red Dye #3. Also be on the lookout for Red #3, Red #40, Yellow #5, Yellow #6, and Blue #1.

POTASSIUM BROMATE

Studies have shown that potassium bromate has a negative impact on Calcium ATPase in the liver, kidneys, and brain.[231,232]

Potassium bromate is an oxidizing agent used in baking, primarily in bread. While providing a strong, well-formed flour mixture, the chemical compound has also been found to have dangerous health side effects, including being a carcinogen.

A report at the NIH states that potassium bromate has been found to be carcinogenic in rats and is dangerous to the kidneys of humans and some lab animals.[233] PubChem, the NIH's library of

231 Farombi E, Alabi M, Akuru T. Kolaviron Modulates Cellular Redox Status and Impairment of Membrane Protein Activities Induced by Potassium Bromate (KBrO3) in Rats. Pharm Res Vol. 45, No 1, 2002.
232 Chuu JJ, Hsu CJ, Lin-Shiau SY. The detrimental effects of potassium bromate and thioglycolate on auditory brainstem response of guinea pigs. Chin J Physiol. 2000 Jun 30; 43(2):91–96.
233 https://www.ncbi.nlm.nih.gov/pmc/articles/PMC1567851/.

medicine, states that potassium bromate has a "Specific target organ toxicity, [even with a] single exposure."[234]

Flour with potassium bromate has been banned in the European Union, China, Canada, and Brazil. However, in the United States, its use is discouraged by the EPA, but not banned.

What can you do? Check the labels. Look for any ingredients that include potassium bromate or bromated flour. Check for it in pastries, crackers, and any variety of bread.

BISPHENOL (BPA)

FOOD PACKAGING INGREDIENT

Studies have shown that bisphenol has a negative effect on Calcium ATPase in the brain, skeletal muscle, and testes.[235,236,237,238]

Bisphenol (BPA) is a chemical compound used in the making of polycarbonate plastic, which is used in food packaging and water bottles. Despite pervasive evidence of the dangers of BPA, it can still be found in the plastic packaging of many foods, the lining of food cans, and kitchen and dental products.

One of the main problems with BPA exposure is that it is an endocrine disruptor, meaning it adversely affects the endocrine system, whose job is to manufacture hormones. These hormones affect so many areas of our bodies, including our reproductive systems, metabolism, growth, and more and are related to chronic states such as obesity and diabetes. In addition, BPA exposure has been linked to abnormalities in the development of fetuses and neonates, increased miscarriages and premature births, increased sperm DNA damage, and, unfortunately, much more.

234 https://pubchem.ncbi.nlm.nih.gov/compound/potassium_bromate.

235 Hughes PJ, McLellan H, Lowes DA, Kahn SZ, Bilmen JG, Tovey SC, Godfrey RE, Michell RH, Kirk CJ, Michelangeli F. Estrogenic alkylphenols induce cell death by inhibiting testis endoplasmic reticulum Ca(2+) pumps. Biochem Biophys Res Commun. 2000 Nov 2; 277(3):568–74.

236 Brown G, Benyon S, Kirk C, Wictome M, East J, Lee A, Michelangeli F. Characterization of a novel Calcium ATPase inhibitor (bis-phenol) and its effects on intracellular Ca2+ mobilization. Biochem Biophy Acta 1994 Nov 2; 1195(2): 252–258.

237 Woeste M, Steller J, Hofmann E, Kidd T, Patel R, Connolly K, Jayasinghe M, Paula S. Structural requirements for inhibitory effects of bisphenols on the activity of the sarco/endoplasmic reticulum calcium ATPase. Bioorg Med Chem. 2013 Jul 1; 21(13):3927–33.

238 Michelangeli F, Tovey S, Lowes DA, Tien RF, Mezna M, McLellan H, Hughes P. Can phenolic plasticising agents affect testicular development by disturbing intracellular calcium homeostasis? Biochem Soc Trans. 1996 May; 24(2):293S.

NONYLPHENOL (NPE)

Studies have shown that nonylphenol has a negative effect on Calcium ATPase in skeletal muscle, adrenal gland, and testes.[239,240,241]

Nonylphenols are often used in plastic food packing. In addition, nonylphenol is used in a wide array of other products, such as detergents. Similar to bisphenol, nonylphenols are endocrine disruptors. Studies have shown that under laboratory conditions, exposure to nonylphenol causes breast cancer cells to proliferate. The substance also has negative effects on growing embryos and newborns. Awareness of this dangerous substance has increased, such that the production and use of nonylphenol is prohibited in the European Union. In the United States, the EPA, concerned about levels in the water supply, is encouraging voluntary reductions in industrial detergents containing NPEs.[242]

GRILLING AND THE CREATION OF BENZO(A)PYRENES

Studies have shown that benzo(a)pyrenes have a negative impact on Calcium ATPase in the brain, red blood cells, and lungs and have an especially deleterious effect on the brain Calcium ATPase during neonatal exposure.[243,244]

Ah, nothing like relaxing over the weekend with a family BBQ. Burgers, steaks, and chicken cooking while you hang with the kids and dog. Couldn't be healthier, right? Well, studies have found that dangers lurk even in the backyard BBQ—specifically through compounds known as benzo(a)pyrenes. The problem occurs when fat from your food drips onto the coals and burns. The burnt fat turns into smoke and rises to coat the food, creating a delicious layer of charred fat and meat. That smoke contains benzo(a)pyrenes.

---

239 Michelangeli F, Ogunbayo O, Wootton L, Lai P, Al-Mousa F, Harris R, Waring R, Kirk C. Endocrine disrupting alkylphenols: structural requirements for their adverse effects on Ca2+ pumps, Ca2+ homeostasis and Sertoli TM4 viability. Chem Biol Interact 2008 Nov 25; 176(2-3):220–26.

240 Liu P, Liu G, Chao W. Effects of nonylphenol on the calcium signal and catecholamine secretion coupled with nicotinic acetylcholine receptors in bovine adrenal chromaffin cells. Toxicology 2008 Feb 3; 244(1):77–85.

241 Logan-Smith M, East J, Lee A. Evidence for a global inhibitor-induced conformation change on the Ca2+-ATPase of sarcoplasmic reticulum from paired inhibitor studies. Biochemistry 2002 Feb 26; 41(8):2869–75.

242 https://en.wikipedia.org/wiki/Nonylphenol.

243 Dong T, Ni J, Wei K, Liang X, Qin Q, Tu B. Effects of benzo(a)pyrene exposure on the ATPase activity and content of Ca2+ in the hippocampus of neonatal SD rats. Zhong Nan Da Xue Xue Bao Yi Xue Ban 2015 Apr; 40(4):356–61.

244 Yang K, Su Q, Qin Q, Tu B. Effect of Benzo(a)pyrene and butylated hydroxylanisole on learning and memory in rats. Sichuan Da Xue Xue Bao Yi Xue-Ban 2016 Jan; 47(1):39–42.

Benzo(a)pyrene exposure has been linked to cancers in the upper gastrointestinal tract, lung, colon, liver, and kidney.[245,246,247] Benzo(a)pyrene has been found by the American Cancer Society to be carcinogenic. Benzo(a)pyrenes have been declared a Group 1 carcinogen (the worst level) by the International Agency for Research on Cancer.[248]

245 https://www.ncbi.nlm.nih.gov/pubmed/11313108.
246 http://www.chicagotribune.com/news/ct-xpm-1986-05-25-8602070649-story.html.
247 https://biofoundations.org/reduce-your-exposure-to-benzopyrenes-as-much-as-possible-then-prevent-their-damage-with-certain-natural-substances/.
248 Ibid.

CHAPTER SIXTEEN

PRESCRIPTION AND OTC DRUGS

There is no doubt that advances in medicine over the decades have saved millions upon millions of lives. However, for the purposes of this chapter and this book, I now delve into some of the unintentional side effects of certain drugs in relation to Calcium ATPase. As with most things, the more aware you are to begin with, the better your decisions will be in the long run.

CELEBREX (CELECOXIB)

Studies have shown that Celebrex inhibits Calcium ATPase in the prostate, smooth muscle, breast, and fibroblasts.[249,250]

You may be familiar with the ever-running Celebrex commercial, "A body in motion tends to stay in motion," showing an older individual bicycling or jumping off a dock into a lake. Celebrex is marketed primarily for arthritis pain, and it seems to work. However, it comes at a price. Celebrex has been shown in numerous cell types to inhibit Calcium ATPase, which may play a role in its cardiovascular risks. In fact, patients who take Celebrex have a 37% increase in major cardiovascular events, including heart attack and stroke. In addition, there

249 Johnson A, Hsu A, Ho-Pi L, Song X, and Chen C. The cyclo-oxygenase-2 inhibitor celecoxib perturbs intracellular calcium by inhibiting endoplasmic reticulum Calcium ATPases; a plausible link with its anti-tumour effect and cardiovascular risks. Biochem J. (2002) 366 831–37.
250 Schonthal A. Direct non-cyclooxygenase-2 targets of celecoxib and their potential relevance for cancer therapy. British Journal of Cancer (2007) 97, 1465–68.

is an 81% increase in gastrointestinal complications, such as bleeding. In 2016, six million prescriptions of Celebrex were filled.[251] Celebrex is part of a class of drugs termed nonsteroidal anti-inflammatory drugs (NSAID), which are common painkillers, and several may be in your drug cabinet today, such as Advil(ibuprofen), or Aleve (naproxen). Although Celebrex is the only one that has been shown to inhibit Calcium ATPase, NSAIDs in general are associated with increased cardiovascular risks.[252] In July 2015, the FDA strengthened the warning that nonaspirin NSAIDs can cause heart attacks or strokes.[253] These negative side effects can occur after only taking the medication for one week, and the higher the dosage, the higher the risk.[254]

TYLENOL (ACETAMINOPHEN)

Most people are familiar with Tylenol, which is used regularly to reduce pain and fever. The active ingredient in Tylenol is *acetaminophen*. Studies have shown that a toxic metabolite of acetaminophen called n-acetyl-p-benzoquinone imine reduces Calcium ATPase in red blood cells and liver tissue, disrupting intracellular calcium regulation leading to cell damage.[255,256,257,258]

ANTIFUNGAL DRUGS: MICONAZOLE, ECONAZOLE, CLOTRIMAZOLE, KETOCONAZOLE

Studies have shown that this class of antifungal drugs (imidazole antifungals) has a negative impact on Calcium ATPase levels in the heart, skeletal muscle, and lymphocytes.[259,260,261,262]

---

251 https://clincalc.com/DrugStats/Top300Drugs.aspx.
252 https://www.webmd.com/osteoarthritis/features/are-nsaids-safe-for-you#1.
253 Food and Drug Administration. 9 July 2015.
254 https://www.bmj.com/content/357/bmj.j1909.
255 Hinds N, Vincenzi N, Differential effects of acylating and oxidizing analogs of N-acetyl-benzoquinoneime on red blood cell membrane proteins. Arch Biochem Biophys. 1990 Nov 15;283(1):200–05.
256 Sub-cellular binding and effects on calcium homeostasis produced by acetaminophen and a nonheptotoxic regioisomer, 3′-hydroxyactanilde in mouse liver> J. Biol Chem. 1989 Jun 15;264(17):9814–19)
257 Nicotera, Rundgren, Porubek, Cotgreave, Moldeus, Orrenius, Nelson. Chem Res Toxicol 1989 Jan-Feb;2 (1):46–50.
258 Moore M, Thor H, Moore G, Nelson S, Moldéus P, Orrenius S. The toxicity of acetaminophen and N-acetyl-p-benzoquinone imine in isolated hepatocytes is associated with thiol depletion and increased cytosolic Ca2+. J Biol Chem. 1985 Oct 25;260(24):13035-40. PMID: 2932433.
259 Mason M, Mayer B, Hymel L. Inhibition of Ca2+ transport pathways in thymic lymphocytes by econazole, miconazole, and SKF 96365. Am J Physiol Cell Physiol 264: C654–62. 1993.
260 Lax A, Soler F, Fernandez-Belda F. Inhibition of sarcoplasmic reticulum Calcium ATPase by miconazole. Am J Physiol Cell Physiol 283; C85–C92. 2002.
261 Snajdrova L, Xu A, Narayanan N. Clotrimazole, and antimycotic drug, inhibits the sarcoplasmic reticulum calcium pump and contractile function in heart muscle. J Biol Chem 1998 Oct 23; 273(43):28032n9.
262 Bartolommei G, Tadini-Buoninsegni F, Hua S, Moncelli M, Inese G, Guidelli R. Clotrimazole inhibits the Calcium ATPase (SERCA) by interfering with the Ca2+ binding and favoring the E2 conformation. J Biol Chem 2006 Apr 7; 281(14):9547–51.

Antifungal drugs are used for yeast infections (Monistat 7), oral thrush, athlete's foot (Lotrimin), toenail fungus, ringworm, and more. The problem in their usage stems from the fact that at a cellular level, many cells in the human body are similar to fungus cells. Therefore, drugs that target fungi have an increased likelihood of hurting healthy human cells. In fact, chronic usage has been associated with dangerous levels of liver toxicity.[263]

Antifungal drugs can be taken orally and in cream form. Although it's unclear how much is absorbed topically, internal application of a cream or gel (as with vaginal use for yeast infections and oral treatments for thrush and nail fungal infections) can potentially cause Calcium ATPase inhibition in a wide range of tissues. Before taking an antifungal drug, think carefully. Most important, get tested to confirm your ugly toenail is the result of an actual fungal infection. According to the American Association of Dermatology, half of suspected cases of fungal infection in nails have a nonfungal cause.[264] Given what we know about the major role reduced Calcium ATPase plays in age related diseases, it may not be a risk worth taking for elderly patients.

ANTIMALARIAL DRUG: MEFLOQUINE

Studies have shown that mefloquine has a negative effect on Calcium ATPase levels in skeletal muscle and the brain.[265,266,267]

The World Health Organization estimates that almost half of the world population is at risk for malaria. Over 200 million cases of malaria occur every year, 400,000 of which lead to death.[268] Given these statistics, malaria drugs play an important role in global health.

One common malaria prophylactic and treatment drug is mefloquine, known under brand names Lariam, Mephaquin, and Mefliam. It is typically taken for one to two weeks before entering an area with malaria. However, in 2013, the FDA issued a warning that

263 https://www.ncbi.nlm.nih.gov/books/NBK8263/.
264 American Academy of Dermatology (February 2013). Five Things Physicians and Patients Should Question. Choosing Wisely: an initiative of the ABIM Foundation. American Academy of Dermatology. Retrieved 2013-12-05.
265 Toovey S. Mefloquine neurotoxcity: a literature review. Travel Medicine and Infectious Disease (2009) 7, 2–6.
266 Toovey S, Bustamante L, Uhlemann A, East J, Krishna S. Effect of Artemisinins and Amino Alchol Partner Antimalarials on Mammalian Sarcoplasmic Reticulum Calcium Adenosine Triphosphatase Activity. Basic and Clinical Pharmacology and Toxicology, 103, 209–13.
267 Dow G, Bauman R, Caridha, Cabezas M, Du F, Gomez-Lobo R, Mefloquine induces dose-related neurological effects in a rat model. Antimicrob Agents Chemother 2006; 50(3);1045–53.
268 http://www.who.int/features/factfiles/malaria/en/.

this drug can have serious neurologic and psychiatric side effects that may persist even after discontinuing use. These side effects include unusual or suicidal behavior, hallucinations, dizziness, and depression.[269] Since this drug inhibits Calcium ATPase in the brain, neurological side effects make sense. If you are taking a trip and need malarial protection, you should speak with your doctor about all the options available.

ALUMINUM HYDROXIDE ANTI-ACIDS

Studies show that aluminum has a negative effect on Calcium ATPase in the brain and skeletal muscles.[270,271,272,273]

Aluminum hydroxide is used for the relief of heartburn and sour stomach. OTC medications that contain aluminum hydroxide include Maalox, Mylanta, Gaviscon, Di-Gel, Acid Gone, and Gelusil. There are calcium-based options available (i.e., Tums).

LIDOCAINE

Lidocaine inhibits Calcium ATPase levels in cardiac[274] and skeletal muscle.[275] Benzocaine has been shown to reduce Calcium ATPase in skeletal muscle.[276]

Topical anesthetics such as lidocaine and benzocaine are medicines that numb and reduce the sensation of pain in the area to which they are applied. In dental and medical procedures, they play a valuable role (imagine a root canal with no numbing!), At the dentist, lidocaine is often injected into the gum or applied as a cream externally. In these situations, lidocaine is applied in

269 https://www.nbcnews.com/healthmain/fda-strengthens-warnings-malaria-drug-6C10783686.
270 Kaur A, Gill KD. Disruption of neuronal calcium homeostasis after chronic aluminium toxicity in rats. Basic Clin Pharmacol Toxicol. 2005 Feb; 96(2):118–22.
271 Sarin S, Julka D, Gill KD. Regional alterations in calcium homeostasis in the primate brain following chronic aluminium exposure. Mol Cell Biochem. 1997Mar;168(1-2):95–100.
272 Julka D, Gill KD. Altered calcium homeostasis: a possible mechanisms of aluminium-induced neurotoxicity. Biochim Biophys Acta. 1996 Jan 17;1315(1):47–54.
273 de Sautu M, Saffioti NA, Ferreira-Gomes MS, Rossi RC, Rossi JPFC, Mangialavori IC. Aluminum inhibits the plasma membrane and sarcoplasmic reticulum Ca(2+)-ATPases by different mechanisms. Biochim Biophys Acta Biomembr. 2018 Aug;1860(8):1580–1588.
274 Karon BS et al. Anesthetics alter the physical and functional properties of the Ca-ATPase in cardiac sarcoplasmic reticulum. Biophysical journal vol. 68,3 (1995): 936–45. doi:10.1016/S0006-3495(95)80269-9.
275 Di Croce DE, Trinks PW, de La Cal C, Sánchez GA, Takara D. Amide-type local anesthetics action on the sarcoplasmic reticulum Ca-ATPase from fast-twitch skeletal muscle. Naunyn Schmiedebergs Arch Pharmacol. 2014 Sep;387(9):873-81. doi: 10.1007/s00210-014-1004-2. Epub 2014 Jun 20. PMID: 24947868.
276 Croce D, Trinks P, Grifo M, Takara D, Sanchez G. Drug action of benzocaine on the sarcoplasmic reticulum Ca-ATPase from fast-twitch skeletal muscle. Naunyn Schmiedebergs Arch Pharmacol 2015 Nov;388(11):1163–70.

measured doses. However, things get murkier with OTC products. They are available as creams, ointments, solutions, eye drops, gels, or sprays and may be applied to areas such as the skin, inside the mouth or throat, in the nose, or in the eyes. Examples include Salon-Pas, Bengay, and Icy Hot. These products are often used on a regular basis, and dosage of creams and sprays is often inexact. The absorption of lidocaine is increased when it is combined with menthol or heat. Skin that is cut or irritated may also absorb more topical medication than healthy skin, so it may be dangerous to use on skin that is raw or blistered, like a severe burn or abrasion. In any situation, a high concentration of lidocaine cream applied over a large area of the body can be absorbed into the bloodstream. This can cause the buildup of toxic levels of lidocaine within the body.

Although generally safe, there have been reports in scientific literature regarding overdoses of lidocaine in a medical setting. Blood levels of lidocaine concentrations were measured in a series of ten patients during and after topical lidocaine spray anesthesia used for a bronchoscopy. Peak serum lidocaine blood concentrations averaged 3.6 micrograms/ml and were attained shortly after the start of the procedure. The researchers concluded that the administration of lidocaine spray may lead to large cumulative doses and blood concentrations that are in the potentially toxic range.[277]

Lidocaine is also a frequently used anesthetic during circumcision procedures. There have been numerous cases in journal articles describing accidental overdoses of lidocaine in infants receiving this treatment.[278,279] Usually treatment with lidocaine is unproblematic, but it is important to adhere to the recommended dose to avoid serious complications. In some cases, a blood disorder called methemoglobinemia can occur, which reduces oxygen levels and if not treated can lead to serious side effects.[280]

277 Labedzki L, Ochs HR, Abernethy DR, Greenblatt DJ. Potentially toxic serum lidocaine concentrations following spray anesthesia for bronchoscopy. Klin Wochenschr. 1983 Apr 1;61(7):379–80. doi: 10.1007/BF01485032. PMID: 6865269.

278 Kjellgard C, Westphal S, Flisberg A. Överdos av prilokain/lidokain kan ge svår methemoglobinemi - Tre fall där krämen gett toxiska effekter hos spädbarn [Intoxication with prilocaine/lidocaine can cause serious methemoglobinemia]. Lakartidningen. 2019 Oct 1;116:FPFT. Swedish. PMID: 31573668.

279 Rezvani M, Finkelstein Y, Verjee Z, Railton C, Koren G. Generalized seizures following topical lidocaine administration during circumcision: establishing causation. Paediatr Drugs. 2007;9(2):125–7. doi: 10.2165/00148581-200709020-00006. PMID: 17407368.

280 Kjellgard C, Westphal S, Flisberg A. Överdos av prilokain/lidokain kan ge svår methemoglobinemi - Tre fall där krämen gett toxiska effekter hos spädbarn [Intoxication with prilocaine/lidocaine can cause serious methemoglobinemia]. Lakartidningen. 2019 Oct 1;116:FPFT. Swedish. PMID: 31573668.

Lidocaine is also used to reduce pain during childbirth. It is important to consider that there is a lidocaine transfer to the infant, as well. In one study, following an epidural, maternal and umbilical levels of lidocaine were detected in thirteen cases, six for caesarean and seven for epidural. The lidocaine given to the mothers crossed to the fetuses readily and resulted in neonatal plasma levels that were half of their mothers'. Levels were tested at three, six, twelve, and twenty-four hours after delivery. It took approximately six hours to reach the half-life of lidocaine. The half-life in infants is at least four times longer than in adults.[281]

A 2018 study reported on an infant who was pale, floppy, and presented with apnea and seizures after birth (similar to Knute's first hours after my epidural). Typical seizure-related conditions such as metabolic disturbances and ischemic encephalopathy were ruled out, and a full range of tests such as MRI, EEG, and spinal tap found nothing. Toxicology screening revealed a toxic concentration of lidocaine in his blood.[282] An earlier French study reported on three similar cases, with two of the cases requiring mechanical ventilation. The good news was that all children recovered.[283]

281 Sakuma S et al. Placental transfer of lidocaine and elimination from newborns following obstetrical epidural and pudendal anesthesia. Pediatric pharmacology (New York, N.Y.) vol. 5,2 (1985): 107–15.

282 Demeulemeester V, Van Hautem H, Cools F, Lefevere J. Transplacental lidocaine intoxication. J Neonatal Perinatal Med. 2018;11(4):439–41. doi: 10.3233/NPM-1791. PMID: 30149475.

283 Pagès H, de la Gastine B, Quedru-Aboane J, Guillemin MG, Lelong-Boulouard V, Guillois B. Intoxication néonatale à la lidocaïne après analgésie par bloc des nerfs honteux: à propos de trois observations [Lidocaine intoxication in newborn following maternal pudendal anesthesia: report of three cases]. J Gynecol Obstet Biol Reprod (Paris). 2008 Jun;37(4):415–18. French. doi: 10.1016/j.jgyn.2008.01.010. Epub 2008 Apr 10. PMID: 18406071.

CHAPTER SEVENTEEN

RECREATIONAL DRUGS

Recreational drugs have been part of human existence from the start. There clearly seems to be some sort of inherent need to alter one's state of consciousness, be it for rituals, community, fun, or simply to escape day-to-day concerns. I have no moral judgments about recreational drugs. My focus is simply on how these substances affect Calcium ATPase and other health-related conditions. The key takeaway: moderation is your best bet.

ALCOHOL

Chronic abuse of alcohol has a negative effect on Calcium ATPase in the heart, brain, skeletal muscle, pancreas, and liver.[284,285,286,287,288,289] In utero, chronic alcohol exposure has a negative effect on Calcium ATPase levels in the heart and brain of the developing fetus.[290,291,292,293]

284 Li Q, Ren J. Cardiac overexpression of metallothionein attenuates chronic alcohol intake-induced cardiomyocyte contractile dysfunction. Cardiovasc Toxicol 2006; (3-4): 173–82.
285 Xia J, Simonyi A, Sun GY. Changes in IP3R1 and SERCA2b mRNA levels in the gerbil brain after chronic ethanol administration and transient cerebral ischemia-reperfusion. Brain Res Mol Brain Res 1998 May; 56(1-2):22–28.
286 Oner P, Cinar F, Kocak H, Gurdol F. Effect of exogenous melatonin on ethanol-induced changes in Na(+), K(+), and Ca(2+)-ATPase activities in rat synaptosomes. Neurochem Res 2002 Dec; 27(12): 1619–23.
287 Rubin E, Katz A, Liever C, Stein E, Puszkin S. Muscle damage produced by chronic alcohol consumption. American Journal of Pathology; Vol 83, No 3 June 1976.
288 González A, Pariente JA, Salido GM. Ethanol impairs calcium homeostasis following CCK-8 stimulation in mouse pancreatic acinar cells. Alcohol. 2008 Nov; 42(7):565–73.
289 Balasubramaniyan V, Viswanathan P, Nalini N. Effect of leptin administration on membrane-bound adenosine triphosphatase activity in ethanol-induced experimental liver toxicity. J Pharm Pharmacol 2006 Aug; 58(8):1113–19.
290 Staley N, Tobin J. Reversible effects of ethanol in utero on cardio sarcoplasmic reticulum of guinea pig offspring. Cardiovasc Res 1991 Jan; 25(1):27–30.
291 Guerri C. Synaptic membrane alterations in rats exposed to alcohol. Alcohol Suppl 1987; 1:467–72.
292 Guerri C, Esquifino A, Sanchis R, Grisolia S. Growth, enzymes, and hormonal changes in offspring of alcohol-fed rats. Ciba Found Symp 1984; 105;85–102.
293 Rudeen PK, Guerri C. The effects of alcohol exposure in utero on acetylcholinesterase, Na/K-ATPase and Ca-ATPase activities in six regions of the brain. Alcohol Alcohol 1985; 20(4):417–25.

That evening glass of wine. Not a problem, right? Or having a few beers after work from time to time? Well, as with so many health-related issues, it depends. While occasional alcohol consumption seems to be safe and, according to the Mayo Clinic, is actually healthy in moderate doses,[294] evidence has been mounting for some time that excessive alcohol consumption is harmful in myriad ways.

First, let's set some parameters for how "moderate" and "heavy" drinking are currently defined. According to the CDC, a drink is defined as twelve ounces of 5% beer, eight ounces of 12% malt liquor, five ounces of 12 % wine, or 1.5 ounces of 40% distilled liquor.[295] The CDC further states that one drink per day for women and two per day for men is considered moderate. Heavy drinking is considered eight or more drinks per week for women and fifteen or more for men. Binge drinking is when four or more drinks for women and five or more for men are consumed in one sitting. Pregnant women or individuals under the age of twenty-one should never drink.[296]

The risks associated with alcohol consumption increase for pregnant women (I also covered this in the neurodevelopment chapter). The CDC states that no amount of alcohol is safe for a pregnant woman, because "When a pregnant woman drinks alcohol, so does her baby."[297] Fetuses are rapidly growing and forming every single day, and alcohol can disrupt the process, sometimes with lifelong neurological and behavioral disabilities. Fetal Alcohol Spectrum Disorder (FASD) is caused by abnormalities in the formation of the brain during this critical time, and children can suffer from the effects for their entire lives. All because their mother drank while pregnant.

The heart, pancreas, liver, skeletal muscle, and brain are all damaged by reduced levels of Calcium ATPase due to alcohol consumption. The risks are higher in utero and can continue all the way through lactation.

---

294 https://www.mayoclinic.org/healthy-lifestyle/nutrition-and-healthy-eating/in-depth/alcohol/art-20044551.
295 https://www.cdc.gov/alcohol/fact-sheets/alcohol-use.htm.
296 https://www.cdc.gov/alcohol/pdfs/alcoholyourhealth.pdf.
297 https://www.cdc.gov/ncbddd/fasd/alcohol-use.html.

TOBACCO

Exposure to cigarette smoke has a negative effect on Calcium ATPase in the brain, sperm, heart, and red blood cells.[298,299,300,301]

The negative health effects of smoking have been common knowledge for years now. Bad breath and stinking clothes aside, according to the CDC, smoking harms every single organ in the human body. One in five deaths in America is related to smoking. 90% of lung cancer cases and 80% of COPD cases directly relate to smoking. The risks for heart disease and stroke are higher, as is the possibility of significant reduction in sperm motility.[302]

Both the nicotine and the nearly four thousand other chemicals in the average cigarette can damage Calcium ATPase levels. The lungs, of course, are affected, but so is cardiovascular function, sperm motility, and mental acuity. As the CDC so clearly stated, smoking damages every single part of your body, and Calcium ATPase levels are no exception.

Cigarettes damage the brain, as well. One study shows that both excessive drinking and excessive smoking have significantly negative effects on memory.[303] Other research shows reduced cognitive abilities due to cigarettes. If a person quits smoking, their abilities improve, but not back to normal levels.[304]

The key takeaway? Don't start smoking, and if you already do, quit now.

---

298 Anbarasi K, Vani G, Balakrishna K, Devi C. Effect of bacoside A on membrane-bound ATPases in the brain of rats exposed to cigarette smoke. J Biochem Mol Toxicol 2005; 19(1):59–65.

299 Kumosani T, Elshal M, Al-Jonaid A, Abduljabar H. The influence of smoking on semen quality, seminal microelements, and Calcium ATPase activity among infertile and fertile men. Clinical Biochemistry 41 (2008) 1199–1203.

300 Pei Z, Zhuang Z, Sang H, Wu Z, Meng R, He E, Scott G, Maris J, Li R, Ren J. Unsaturated aldehyde crotonaldehyde triggers cardiomyocyte contractile dysfunction: role of TRPV1 and mitochondrial function. Pharmaol Res 2014 Apr; 82;40–50.

301 Chen Z, Jiao X, Yang F, Wu H, Shu H. Effect of huoxuequyu recipe on erythrocyte ultrastructure and membrane ATPase activity in rats with passive smoking. J Tongji Med Univ 1996; 16(4): 220–22.

302 https://www.cdc.gov/tobacco/data_statistics/fact_sheets/health_effects/effects_cig_smoking/index.htm.

303 https://www.frontiersin.org/articles/10.3389/fpsyt.2016.00075/full.

304 https://www.newsweek.com/smoking-brain-smokers-memory-learning-mind-body-525347.

COCAINE

Studies show that Cocaine has a negative effect on Calcium ATPase in the heart and arteries.[305,306]

Cocaine? That's a leftover vice from the 1980s Age of Excess, right? Red Corvettes, gold chains, rising stock markets, and cocaine! Unfortunately, that's not the whole story. Today, over 1.5 million people in the US alone use cocaine every month.[307]

One of the primary risks of cocaine is within the cardiovascular system. This is where Calcium ATPase plays a distinctive role. As you've already read, reduced Calcium ATPase levels in heart muscle can have devastating effects. Cocaine is the leading drug-related reason for trips to the emergency room, most of which are related to chest pain.[308] So cocaine itself stresses the heart, but with the addition of its deleterious effect on Calcium ATPase? A double whammy.

Chronic cocaine use can damage the gastrointestinal tract and has also been linked to an increased risk of stroke and aortic ruptures. When it comes to the brain, chronic use can cause bleeding and increased cognitive dysfunction. What does this look like? Attention deficits, memory loss, difficulty making decisions, and even Parkinson's.[309]

In addition, evidence shows that repeated cocaine use actually rewires the brain, creating a vicious cycle of needing higher highs while suffering lower lows. In extreme cases, users can actually detach from reality, suffering psychosis and hallucinations.[310]

---

305 Wang J, Yan X, Min J, Sullivan M, Hampton T, Morgan J. Cocaine downregulates cardiac SERCA2a and depresses myocardial function in the murine model. Can J. Physiol Pharmacol Vol. 80, 2002.

306 Togna G, Graziani M, Russo P, Caprino L. Cocaine toxic effect on endothelium-dependent vasorelaxation: an in vitro study on rabbit aorta. Toxiocol Letters 123 (2001) 43–50.

307 https://skywoodrecovery.com/cocaine-addiction-statistics/.

308 https://www.acc.org/latest-in-cardiology/ten-points-to-remember/2017/06/27/13/58/the-cardiovascular-effects-of-cocaine.

309 Ibid.

310 https://www.drugabuse.gov/publications/research-reports/cocaine/what-are-long-term-effects-cocaine-use.

MARIJUANA

Marijuana can have a negative effect on Calcium ATPase levels in the brain,[311] the heart,[312] the skeletal muscle,[313] and male reproductive organs.[314]

The main psychoactive chemical in marijuana is THC (delta-9-tetrahydrocannabinol). Marijuana's effect is correlated to the amount and potency of the THC it contains. Although there are definitely some desirable effects of THC, including pain relief, stress reduction, and improved sleep, there is a potential downside to chronic use. Interestingly, these negative side effects are found in areas of the body where THC has been reported to reduce the level of Calcium ATPase. Cardiovascular effects can include acute coronary syndrome, coronary thrombosis, myocardial infarction, cardiomyopathies, heart failure, stroke, vasospasm, vascular inflammation, atrial fibrillation, heart block, and sudden death.[315] Negative cognitive effects of chronic use can accumulate over a lifetime. As one researcher puts it, "a large body of convergent data suggest that long term use of marijuana may cause significant abridgement of one's potential." It also reduces sperm concentration. A 2018 study conducted at Duke University that involved 24 human subjects found that cannabinoid use reduced sperm concentration approximately 40%.[316]

311 Bloom AS, Haavik CO, Strehlow D. Effects of delta9-tetrahydrocannabinol on ATPases in mouse brain subcellular fractions. *Life Sci*. 1978;23(13):1399–1404.

312 Al Kury LT, Voitychuk OI, Ali RM, et al. Effects of endogenous cannabinoid anandamide on excitation-contraction coupling in rat ventricular myocytes. *Cell Calcium*. 2014;55(2):104–18.

313 Oláh T, Bodnár D, Tóth A, et al. Cannabinoid signalling inhibits sarcoplasmic Ca^{2+} release and regulates excitation-contraction coupling in mammalian skeletal muscle. *J Physiol*. 2016;594(24):7381–7398. doi:10.1113/JP272449.

314 Dalterio SL, Bernard SA, Esquivel CR. Acute delta 9-tetrahydrocannabinol exposure alters Calcium ATPase activity in neuroendocrine and gonadal tissues in mice. *Eur J Pharmacol*. 1987;137(1):91–100. doi:10.1016/0014-2999(87)90186-5.

315 Pacher P, Steffens S, Haskó G, Schindler TH, Kunos G. Cardiovascular effects of marijuana and synthetic cannabinoids: the good, the bad, and the ugly. *Nat Rev Cardiol*. 2018;15(3):151–66.

316 Murphy SK, Itchon-Ramos N, Visco Z, et al. Cannabinoid exposure and altered DNA methylation in rat and human sperm. *Epigenetics*. 2018;13(12):1208–21.

CHAPTER EIGHTEEN

STRESS

Most of us are aware that stress is bad for your heart. It turns out that one reason is the negative effect stress has on Calcium ATPase. We all have stress in our lives, and a little bit of it can be good for us. The problem arises when we have chronic stress, the type that becomes a habit. We all know this kind of stress is bad for us, but let's look at it in relation to Calcium ATPase.

Chronic stress is bad for our health because it has a negative effect on Calcium ATPase in the heart. There is a correlation between the stress hormones called *catecholamines* and reduced Calcium ATPase. Catecholamines are hormones made by your brain, nerve tissues, and adrenal glands in response to emotional or physical trauma. They are responsible for the fight-or-flight response triggered by fear (whether real or imaginary). Examples include epinephrine (adrenaline) and norepinephrine (noradrenaline). Once released, these hormones increase your heart rate and blood pressure in order to send more blood to your major organs, such as your brain and heart, and cause you to breathe faster to provide more oxygen to your cells.

If elevated for extended periods of time, high catecholamine levels in blood reduce Calcium ATPase. Researchers examined the effects of exposing rat cardiac muscle cells to norepinephrine. Under laboratory conditions, forty-eight hours of exposure to norepinephrine resulted in a reduction of Calcium ATPase by 56%.[317]

317 Satoh N, Suter T, Liao R, Colucci W. Chronic alpha-adrenergic receptor stimulation modulates the contractile phenotype of cardiac myocytes in vitro. Circulation 2000 Oct 31; 102(18):2249–54.

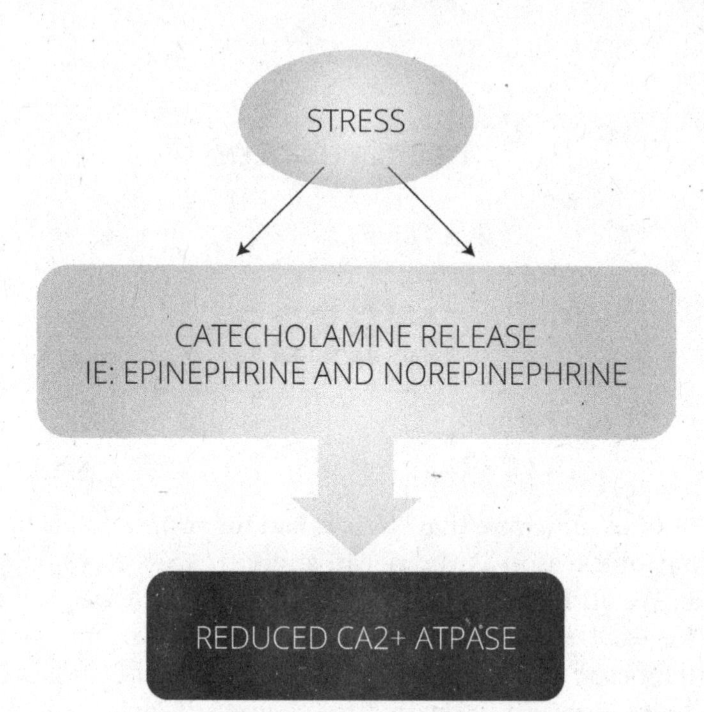

Another study examined the effect of social stress on Calcium ATPase levels in the heart. Chronic social stress was induced by setting up a mix of male and female rats in close proximity housing. In such situations, hierarchies of male dominance develop, and subordinate rats can be identified through various stress markers, such as submissive behaviors, wound patterns, and reduced body weight. The salient fact for this particular study is that this design closely resembles the loss of social control in humans. In line with this reasoning, the rats at the bottom of the social ladder were identified as the "stressed group," and their Calcium ATPase levels were compared to a control group of rats housed in normal conditions. The results were astonishing. The rats who experienced social stress had 40% lower Calcium ATPase levels in the heart than the animals in the control group. No surprise: the stressed group with reduced Calcium ATPase had cardiac problems that mimicked heart failure.[318]

318 Turdi S, Yuan M, Leedy G, Wu Z, Ren J. Chronic Social Stress Induces Cardiomyocyte Contractile Dysfunction and Intracellular Ca2+ Derangement in Rats. Physiol Behav 2012 Jan 18; 105(2):498–509.

Another set of researchers also studied the connection between stress and Calcium ATPase, but in the context of broken heart syndrome (*stress cardiomyopathy* in medical language). Broken heart syndrome is a reaction of an individual that disrupts the functioning of a portion of the heart muscle and leads to acute heart failure. The condition is triggered by emotionally stressful events (such as the death of a spouse). Fortunately, if the condition is treated promptly, it reverses itself in several days or weeks.

Researchers measured the Calcium ATPase levels of ten patients experiencing broken heart syndrome and compared these levels with normal hearts that served as the control. Calcium ATPase activity was significantly reduced in broken heart syndrome patients. The researchers concluded that this reduction in Calcium ATPase activity plays a role in cardiac dysfunction during stress cardiomyopathy.[319]

In addition, an animal study compared Calcium ATPase levels in rats with simulated stress cardiomyopathy to Calcium ATPase levels in a control group. They found a significant reduction in Calcium ATPase activity in the stress cardiomyopathy rats and concluded that reduced Calcium ATPase levels in stress cardiomyopathy contribute to cardiac dysfunction.[320]

In summary, stress causes the release of *catecholamines,* which have a negative impact on Calcium ATPase levels. In the program section, we give you a plan to reduce catecholamines and protect your Calcium ATPase.

---

319 Nef H, Mollma H, Trodi C, Kostin S, Voss S, Hilpert P, Behrens C, Rolf A, Rixe J, Weber M, Hamm C, Elasser A. Abnormalities in intracellular Ca2+ regulation contribute to the pathomechanism of Tako-Tsubo cardiomyopathy. Eur Heart J 2009 Sep; 30(17):2155–64.

320 Hata T, Hozaima L, Sandhu M, Panagia V, Dhalla N. Role of oxidative stress in catechoaminhe-induced changes in cardiac sarcolemmal Ca2+ transport.Tappia P, Arch Biochem Biophys 2001 Mar 1; 387(1):85–92.

SECTION 4

The Calcium ATPase Max Protocol (CAMP)

Here is the most important part of the book—the actions you can take to maximize your Calcium ATPase levels. I call it the **Calcium ATPase Max Protocol or CAMP,** and it consists of four distinct Plans that work synergistically to support optimal health:

1. **Toxin-Reduction Plan**
2. **Flexible Food and Nutrition Plan**
3. **Targeted Exercise Plan**
4. **Stress Management Plan**

Each Plan noted above comes with a set of Guiding Principles that I outline in the following chapters. Likely CAMP dovetails into the positive actions you are already taking. In which case, you can glean from this what you wish and add it to your ongoing practices with the understanding that perhaps you have new insight into the intracellular health you're generating. But I will guarantee you this: if all you do is follow the CAMP Protocol, you will be covering most of your bases for optimal health.

CHAPTER NINETEEN

CAMP TOXIN-REDUCTION PLAN

PRINCIPLE ONE	Clean Air
PRINCIPLE TWO	Clean Water
PRINCIPLE THREE	Clean Food

Now that you are aware (or have been reminded) of the toxins in our environment, living cleanly may seem a bit daunting. Where do I begin? Start with the basics: clean air, clean water, and clean food. After you have the basics addressed, you can start looking at the nitty-gritty. At the end of this chapter, I have provided detailed checklists for addressing specific sources of toxins in your home and environment. I suggest you take these one at a time, understand the risks and where they hide, and give yourself the space to do your best to mitigate them. If you can't address all issues at once with regard to a particular toxin, that's okay: just make a note of it at the bottom of the checklist for that toxin. Aim for one or two every month. It may take you several rounds to address all suggestions, or you may not be able to address them all. Just take one step at a time and appreciate the fact that every small step you take helps reduce you and your family's exposure to Calcium ATPase inhibitors. That alone is a huge step toward better health.

PRINCIPLE ONE
Clean Air

The Science

Dust can contain a host of environmental toxins, including lead, pesticides, and fire retardants.[321,322,323,324,325] Young children ingest more dust than adults[326] and are up to ten times more vulnerable to such exposure. Early contact with pollutants found in dust is associated with higher rates of asthma, loss of intelligence, ADHD, and cancer.[327,328]

The Action: Clean It Up

- Upgrade your furnace and cooling system filters with an electrostatic filter. Replace the filter every 1–3 months.
- Consider buying a HEPA stand-alone air filter. I recommended AirDoc.
- The most valuable item in reducing dust is a vacuum with a HEPA filter.
- Wet-mop floors once or twice a week.
- Use microfiber cloths to dust surfaces. Wet cloths work best, so when you're dusting furniture that isn't wooden, try wetting the cloth first. If you have small children, make sure to wipe down toys and play areas where dust accumulates.

---

321 Aslam I, Mumtaz M, Qadir A, Jamil N, Baqar M, Mahmood A, Ahmad SR, Zhang G. Organochlorine pesticides (OCPs) in air-conditioner filter dust of indoor urban setting: Implication for health risk in a developing country. Indoor Air. 2020 Nov 28. doi: 10.1111/ina.12772. Epub ahead of print. PMID: 33247439.

322 Kassotis CD, Hoffman K, Phillips AL, Zhang S, Cooper EM, Webster TF, Stapleton HM. Characterization of adipogenic, PPARγ, and TRγ activities in house dust extracts and their associations with organic contaminants. Sci Total Environ. 2020 Nov 14:143707. doi: 10.1016/j.scitotenv.2020.143707. Epub ahead of print. PMID: 33223163.

323 Whitehead TP, Crispo Smith S, Park JS, Petreas MX, Rappaport SM, Metayer C. Concentrations of Persistent Organic Pollutants in California Children's Whole Blood and Residential Dust. Environ Sci Technol. 2015 Aug 4;49(15):9331–40. doi: 10.1021/acs.est.5b02078. Epub 2015 Jul 22. PMID: 26147951.

324 Roberts JW, Dickey P. Exposure of children to pollutants in house dust and indoor air. Rev Environ Contam Toxicol. 1995;143:59–78. doi: 10.1007/978-1-4612-2542-3_3. PMID: 7501867.

325 Ertl H, Butte W. Bioaccessibility of pesticides and polychlorinated biphenyls from house dust: in-vitro methods and human exposure assessment. J Expo Sci Environ Epidemiol. 2012 Nov;22(6):574–83. doi: 10.1038/jes.2012.50. Epub 2012 Jun 13. PMID: 22692365.

326 Zhou J, Mainelis G, Weisel CP. Pyrethroid levels in toddlers' breathing zone following a simulated indoor pesticide spray. J Expo Sci Environ Epidemiol. 2019 Apr;29(3):389-96. doi: 10.1038/s41370-018-0065-6. Epub 2018 Sep 5. PMID: 30185948; PMCID: PMC7323485.

327 Kadawathagedara M, de Lauzon-Guillain B, Botton J. Environmental contaminants and child's growth. J Dev Orig Health Dis. 2018 Dec;9(6):632-41. doi: 10.1017/S2040174418000995. PMID: 30720417.

328 Jurewicz J, Hanke W, Johansson C, Lundqvist C, Ceccatelli S, van den Hazel P, Saunders M, Zetterström R. Adverse health effects of children's exposure to pesticides: what do we really know and what can be done about it. Acta Paediatr Suppl. 2006 Oct;95(453):71–80. doi: 10.1080/08035320600886489. PMID: 17000573.

- Wash sheets and pillowcases weekly. Other items, such as comforters and blankets, can be washed monthly. For nonwashable items, a good shake outside can help eliminate dust buildup.
- Every few months, eliminate as much dust as you can from cushions and rugs. Take them outside and beat them all over with a broom handle until you no longer see dust particles released into the air. Also, make sure to vacuum underneath the cushions and rugs when they are removed.
- Do not use indoor chemical pesticides. If you have a pest control issue, look for natural options. Even big box stores now carry less toxic pest control. There are also pest control companies that provide this service.
- Avoid tracking pesticides and toxins indoors. Mud and dirt brought into the house on the bottom of your shoes contribute to household dust.
- Groom your pets often and wash their bedding frequently.
- Do not use chemical-laden air fresheners in the house.
- Use fewer toxic cleaning supplies readily available (even at the local CVS!).
- Use caulk to seal cracks around door and window frames.

PRINCIPLE TWO
Clean Water

The Science

Without water we die. What's important to know for our purposes is that the water you drink may be a cocktail of Calcium ATPase inhibitors. Lead, mercury, cadmium, chlorine, atrazine, fluoride, bisphenol, nonylphenol, and a variety of other toxins and pesticides are commonly found in tap water.[329,330,331,332,333,334,335,336,337,338,339]

Don't think of bottled water as the solution. In addition to its heavy price tag and the negative environmental impact, bottled water is no guaranteed protection from toxins.[340] In fact, a study done by the Environmental Working Group found that industrial chemicals, disinfection byproducts, prescription drugs, and even bacteria can be found in bottled waters.[341] Furthermore, even BPA-free plastic bottles are still just that—plastic. Why drink water that has been stored (in all likelihood for several months) in an airtight plastic container?

---

329 Tomlinson MS, Bommarito P, George A, Yelton S, Cable P, Coyte R, Karr J, Vengosh A, Gray KM, Fry RC. Assessment of inorganic contamination of private wells and demonstration of effective filter-based reduction: A pilot-study in Stokes County, North Carolina. Environ Res. 2019 Oct;177:108618. doi: 10.1016/j.envres.2019.108618. Epub 2019 Aug 2. PMID: 31419714; PMCID: PMC6717535.

330 Bexfield LM, Belitz K, Lindsey BD, Toccalino PL, Nowell LH. Pesticides and Pesticide Degradates in Groundwater Used for Public Supply across the United States: Occurrence and Human-Health Context. Environ Sci Technol. 2020 Dec 14. doi: 10.1021/acs.est.0c05793. Epub ahead of print. PMID: 33315392.

331 Saleeby B, Shimizu MS, Sanchez Garcia RI, Avery GB, Kieber RJ, Mead RN, Skrabal SA. Isomers of emerging per- and polyfluoroalkyl substances in water and sediment from the Cape Fear River, North Carolina, USA. Chemosphere. 2021 Jan;262:128359. doi: 10.1016/j.chemosphere.2020.128359. Epub 2020 Sep 16. PMID: 33182107.

332 Padhye LP, Yao H, Kung'u FT, Huang CH. Year-long evaluation on the occurrence and fate of pharmaceuticals, personal care products, and endocrine disrupting chemicals in an urban drinking water treatment plant. Water Res. 2014 Mar 15;51:266–76. doi: 10.1016/j.watres.2013.10.070. Epub 2013 Nov 7. PMID: 24262763.

333 Sanders AP, Desrosiers TA, Warren JL, Herring AH, Enright D, Olshan AF, Meyer RE, Fry RC. Association between arsenic, cadmium, manganese, and lead levels in private wells and birth defects prevalence in North Carolina: a semi-ecologic study. BMC Public Health. 2014 Sep 15;14:955. doi: 10.1186/1471-2458-14-955. PMID: 25224535; PMCID: PMC4190372.

334 Evans S, Campbell C, Naidenko OV. Analysis of Cumulative Cancer Risk Associated with Disinfection Byproducts in United States Drinking Water. Int J Environ Res Public Health. 2020 Mar 24;17(6):2149. doi: 10.3390/ijerph17062149. PMID: 32213849; PMCID: PMC7142415.

335 Bexfield LM, Toccalino PL, Belitz K, Foreman WT, Furlong ET. Hormones and Pharmaceuticals in Groundwater Used As a Source of Drinking Water Across the United States. Environ Sci Technol. 2019 Mar 19;53(6):2950–60. doi: 10.1021/acs.est.8b05592. Epub 2019 Mar 5. PMID: 30834750.

336 Fu J, Lee WN, Coleman C, Nowack K, Carter J, Huang CH. Removal of pharmaceuticals and personal care products by two-stage biofiltration for drinking water treatment. Sci Total Environ. 2019 May 10;664:240–48. doi: 10.1016/j.scitotenv.2019.02.026. Epub 2019 Feb 2. PMID: 30743118.

337 Chowdhury S. Exposure assessment for trihalomethanes in municipal drinking water and risk reduction strategy. Sci Total Environ. 2013 Oct 1;463-464:922–30. doi: 10.1016/j.scitotenv.2013.06.104. Epub 2013 Jul 19. PMID: 23872246.

338 Klecka G, Persoon C, Currie R. Chemicals of emerging concern in the Great Lakes Basin: an analysis of environmental exposures. Rev Environ Contam Toxicol. 2010;207:1-93. doi: 10.1007/978-1-4419-6406-9_1. PMID: 20652664.

339 Stackelberg PE, Furlong ET, Meyer MT, Zaugg SD, Henderson AK, Reissman DB. Persistence of pharmaceutical compounds and other organic wastewater contaminants in a conventional drinking-water-treatment plant. Sci Total Environ. 2004 Aug 15;329(1–3):99-113. doi: 10.1016/j.scitotenv.2004.03.015. PMID: 15262161.

340 https://www.consumerreports.org/bottled-water/whats-really-in-your-bottled-water/.

341 Le HH, Carlson EM, Chua JP, Belcher SM. Bisphenol A is released from polycarbonate drinking bottles and mimics the neurotoxic actions of estrogen in developing cerebellar neurons. Toxicol Lett. 2008 Jan 30;176(2):149–56. doi: 10.1016/j.toxlet.2007.11.001. Epub 2007 Nov 19. PMID: 18155859; PMCID: PMC2254523.

The Action: Clean It Up

- Opt into a *reverse osmosis filter*. It's the best type of filter for eliminating toxins, and my preference is Aquatrue. This is the first osmosis filter that sits on your countertop rather than requiring a plumber for setup.
- If reverse osmosis is not an option, activated carbon filters are better than nothing, but they're not all created equal. For example, if you buy Brita brand, make sure you get the special long-lasting filter that filters out lead. The Pur special filter removes lead and some pesticides, but there are other toxins it doesn't remove.
- Do your homework on water filters. Manufacturers are required to list the toxins removed by their systems.
- If bottled water is your thing, a healthier option for you and the environment is buying a stainless steel bottle for you and your family members (get a different color for each person); you can have several on hand in the fridge prefilled with filtered water to make things easier.

PRINCIPLE THREE
Clean Food

The Science

I covered the negative impact of food additives and preservatives on Calcium ATPase in Section Two: The Enemies of Calcium ATPase. (I know it is not light reading, but feel free to read again if you need motivation to make changes!) Below, I cover how to go about cleaning up your food sources for optimal Calcium ATPase health.

The Action: Clean It Up

While I surely can't tackle every single way I would like you to clean up your food, I will offer a tip-of-the-iceberg view into the actions you can take most immediately. When it comes to the foods you choose, please:

1. Avoid ultraprocessed foods, specifically those with chemicals and dyes.
2. Reduce consumption of sweets and high carbohydrate foods.
3. Consume whole fresh foods, mostly organic if possible.
4. Avoid consumption of seafood with high mercury levels.

#1 Avoid Ultraprocessed Foods (look out for TBHQ, BHA/BHT, Food Dyes, and Aluminum)

Read ingredient lists when purchasing packaged foods and make sure that you recognize every ingredient in your choice. I have listed some examples below in terms of the everyday foods that contain these additives.

TBHQ (preservative)	Reese's Peanut Butter Cups, Snickers, Nestle Crunch, Butterfinger, Dove chocolate, Ghirardelli chocolate, Kellogg's Rice Krispies Treats, Kellogg's Pop-Tarts, Cheez-It, Tastykakes, Lil Debbie products including Nutty Bars, Swiss Rolls, and Oatmeal Creme Pie (my favorite growing up!)
BHA/BHT (preservative)	**Cereals:** Frosted Flakes, Frosted Mini-Wheats, Honey Bunches of Oats, Rice Krispies, Fruit Loops, Raisin Bran **Snacks:** Triscuit, Wheat Thin, Keebler graham cracker sticks, YoCrunch low-fat yogurt with granola **Frozen Food:** DiGiorno frozen pizza, Tyson chicken breast tenderloins, Banquet chicken pot pies **Gum:** Orbit, Trident, Wrigley

Food Dyes	**Frostings:** Betty Crocker, Duncan Hines, Pillsbury frostings, cake gels, colored sugars, McCormick food dyes **Fruit cocktails:** Dole fruit bowls, Libby fruit cocktail **Candy:** Pink and blue marshmallow Peeps, Salt-water taffy, Colored rock candy **Medications:** Bubble gum flavored antibiotics, Pediasure strawberry flavored kids' nutrition drink
Aluminum	Eggo waffles, Pillsbury refrigerated sugar cookie dough, Nestle Tollhouse refrigerated cookie dough, Hostess Twinkies and Donettes, McDonald's chicken nuggets, Pillsbury biscuits and crescent rolls, Bisquick and Aunt Jemima pancake mix, Pop-Tarts, DiGiorno pizza, Totino's pizza rolls, Jimmy Dean's sausage biscuits, Betty Crocker and Duncan Hines cake mix

#2 Reduce Consumption of Sweets and High Carbohydrate Foods

To avoid blood sugar surges, combine sweets with a protein or fat to slow down their absorption. Of course, there will be the times when you walk to the ice cream store as a family and savor the moment, or other special occasions such as birthdays and Halloween. You just want to be cognizant of avoiding sugar surges as an everyday event. Sugary foods are often consumed as snacks. And there is a way to swap the sugar for something more nutritious and even delicious! Here are some go-to snacks that I have called on over the years, and trust me when I say that taking a few extra minutes to ensure that you and your family have nourishing noshes makes all the difference. It's also worth the effort in the long run! Buy some snack-sized containers and dedicate a shelf in the refrigerator and or pantry everyone knows where to look:

- Raw almonds and a few dark chocolate chips
- Carrot sticks with hummus
- Organic string cheese and organic rice crackers with sliced Roma tomatoes
- Fresh apple circles with nut butter and a few chocolate chips on top
- Kale chips with guacamole

The CAMP Food and Nutrition Plan, in the following chapter, will provide a more extensive meal and snack guide, so this is just a quick start.

#3 Consume whole fresh foods, mostly organic if possible

Data from the Food and Drug Administration Pesticide Residue Monitoring Program showed that in 2016, approximately 47% of domestic foods and 49% of imported foods sampled had detectable pesticide residue. Biomonitoring data from 1999–2002 showed metabolites from pyrethroid insecticides in 70% of urine samples. So yes, pesticides from food really do end up in our bodies.

The good news is that numerous studies have shown that adopting an organic diet can result in substantially lower urinary pesticide levels. Organophosphate pesticides metabolites were reduced by up to 96% after a week of an organic diet.[342]Metabolites of pyrethroid pesticides were also reduced significantly after one week of an organic diet.[343] Both of these pesticides are inhibitors of Calcium ATPase. So in short, an organic diet is worth it!

The Environmental Working Group (EWG) is an incredible organization that provides a list of foods that have the greatest level of pesticides based on government data. Avoid these foods unless organic. If fresh organic is too expensive, consider frozen organic foods, which are typically less expensive and still have nutritional value, as they are flash frozen at the peak of freshness.

342 Oates L, Cohen M, Braun L, Schembri A, Taskova R. Reduction in urinary organophosphate pesticide metabolites in adults after a week-long organic diet. Environ Res. 2014 Jul;132:105-11. doi: 10.1016/j.envres.2014.03.021. Epub 2014 Apr 25. PMID: 24769399.

343 Lu C, Toepel K, Irish R, Fenske RA, Barr DB, Bravo R. Organic diets significantly lower children's dietary exposure to organophosphorus pesticides. Environ Health Perspect. 2006 Feb;114(2):260-63. doi: 10.1289/ehp.8418. PMID: 16451864; PMCID: PMC1367841.

EWG 2020 Dirty Dozen Foods List

The Dirty Dozen is a list of twelve fruits and vegetables that contain the highest concentrations of pesticides after being washed. Since organic produce is more expensive than conventional produce, and many people cannot afford to purchase organic fruits and veggies exclusively, the EWG recommends reserving your organic grocery budget for when you buy items that are part of the Dirty Dozen.

STRAWBERRIES	SPINACH	KALE
NECTARINES	APPLES	GRAPES
PEACHES	CHERRIES	PEARS
TOMATOES	CELERY	POTATOES

The EWG also published a 2020 Dirty Dozen PLUS™ list based on recent research findings. Pay attention to the following:

- **Sweet bell peppers and hot peppers**: Both can carry residues of neurotoxic chemicals such as the organophosphate insecticides chlorpyrifos.
- **Oats**: Recent studies by EWG have detected the toxic pesticide glyphosate in more than 95 percent of samples of oat-based products, including children's cereals.
- **Beans and Legumes**: Testing by the Canadian Food Inspection Agency detected glyphosate in nearly half of bean, pea, and lentil products tested in 2015 and 2016. To avoid glyphosate in these products, organic varieties are a good choice.
- **Herbs**: Certain herbs such as cilantro can be contaminated with pesticide residue such as Calcium ATPase inhibitors chlorpyrifos and pyrethroids. The pesticide levels can be as high as those found on conventionally grown spinach and kale, both of which are on the EWG's Dirty Dozen list.

- **Rice and Wheat**: In 2018, USDA collected and tested 758 wheat flour and 189 rice samples and found 19 and 37 different pesticides on these commodities, respectively, including the neurotoxic pesticide chlorpyrifos and the pyrethroid insecticide deltamethrin.

  In addition, a recent study in France found that children who consumed greater amounts of pasta, rice, semolina, breakfast cereals, and whole grain bread had higher levels of metabolites of pyrethroid pesticides in their urine compared to those who consumed less of these foods.
- **Raisins**: EWG experts say that 99% of nearly 700 raisin samples tested positive for traces of at least two pesticides, with one sample containing 26 different pesticides. "If we included raisins in our calculations, they would be number one on the Dirty Dozen," said Thomas Galligan, PhD, a toxicologist for the EWG, in a press release.

EWG 2020 Clean Fifteen Food List

Here is some good news! These conventionally raised produce items had the lowest amount of pesticide residues, with nearly all of them containing only four or fewer pesticides. Almost 70% of these samples had no pesticide residues whatsoever.

AVOCADOS	SWEET CORN	PINEAPPLE
ONIONS	PAPAYA	SWEET PEAS (FROZEN)
EGGPLANT	ASPARAGUS	CAULIFLOWER
CANTALOUPES	BROCCOLI	MUSHROOMS
CABBAGE	HONEYDEW MELON	KIWI

#4 Avoid consumption of seafood with high mercury levels

Mercury is a Calcium ATPase inhibitor, and kids are especially vulnerable to its negative effects. However, fish is a great source of important nutrients such as DHA. Don't give up eating fish; just be careful to make the safest choices.

Tuna has a low level of mercury, and "light" is best. The EPA and FDA rank canned light tuna (solid or chunk) to be among the best choices for children to eat, recommending 2–3 servings a week. "Light" tuna, which means it has a pinkish color, includes species such as skipjack. This is considered a better choice than white (albacore) and yellowfin tuna, although these are still considered good choices.

Salmon, trout, and herring are also considered low in mercury and high in brain-boosting DHA. Other types of seafood considered to be good choices include shrimp, cod, catfish, crab, scallops, pollock, tilapia, whitefish, perch, flounder, sole, sardine, anchovy, crawfish, clams, oyster, and lobster.

SEAFOOD TO AVOID COMPLETELY:		
SWORDFISH	SUSHI-GRADE TUNA	TILEFISH
SHARK	KING MACKEREL	ORANGE ROUGHY
MARLIN	BIGEYE TUNA	BLUEFIN TUNA

One note about sushi: Sushi-grade tuna is a very large fish and contains high levels of mercury. Steer your kids away from tuna sushi.

OTHER CONSUMPTION CONSIDERATIONS PARTICULARLY FOR CHILDREN

Because being a mom is my life's work and what led me to write this book, I want to offer a special note to parents. I know parenting under normal circumstances is not easy. However, I think once you gain the knowledge you need to make healthy choices, it will not be too difficult, and you will get the benefit of knowing you are taking great care of your kid's health.

In an ideal world:

- All children should be protected as much as possible from Calcium ATPase inhibitors in their environment and diet.
- All children's diets should include nutrient-dense food that supports Calcium ATPase.
- All children should believe that they matter. This is not a war. This is a team effort.

Here are some specific items you should be keenly aware of as a parent

#1 Pay attention to how your kids use toothpaste

Dentists now recommend children only use a pea-sized amount of toothpaste rather than a strip covering the length of the brush. Supervise brushing for younger children. Make sure toothpaste is rinsed out and not swallowed. Avoid candy-flavored toothpaste. Be cautious about using fluoridated toothpaste in children younger than age six, as they are more likely to swallow toothpaste instead of spitting it out.

Keep toothpaste out of reach of children. If a child ingests one tube of toothpaste—tempting when candy-flavored—it is enough to cause fluoride poisoning. That is why the FDA requires the following warning label on every toothpaste that contains fluoride: "Warning: Keep out of the reach of children under six years of age. If you accidentally swallow more than used for brushing, seek professional assistance or contact a Poison Control Center immediately."

Also, reduce intake of foods made with mechanically separated meat (e.g., chicken fingers, nuggets), especially in young children. These products contain elevated levels of fluoride due to the contamination from bone particles that occurs during the mechanical deboning process. This is especially relevant in products such as infant food. A single serving of infant food made with mechanically deboned chicken contains 87% of the upper limit for safety of fluoride. For a one-year-old toddler, a single serving of chicken sticks (71g) provides 50% of the upper limit for safety of fluoride.

#2 Pay attention to medications you give your children

Choose dye-free when that is an option. For example, choose dye-free Pediolyte, Benedryl, cough drops, and vitamins. Avoid popular over-the-counter products containing benzocaine such as Chloraseptic throat spray, Cepacol cough drops, and Vicks Vapo Cool throat drops.

Avoid chemical lice products. Listen to your doctor but pay attention to your gut. If you notice a negative side effect, reach out—maybe it's common for that product, or maybe there is a different option.

#3 Make smart choices when choosing sunscreens

Avoid chemical sunscreen. Common chemical sunscreen ingredients to avoid include oxybenzone, avobenzone, octisalate, octocrylene, homosalate, and octinoxate. Avoid all sunscreen sprays. Choose non-nanoparticle titanium dioxide and zinc oxide brands. Some examples are Badger-Organic Baby Sunscreen SPF 30, All Good-Organic Sunscreen for Kids SPF 30, and Nurture My Body-Unscented Baby Organic Sunscreen SPF 30.

I have left most of the personal care products issues on the sideline because the amount of chemical present in everyday products is mind-bending. What I suggest is that you start with one product at a time, e.g., sunscreen. Perhaps next you look for less chemically laced body soap, and so on through the personal products (including makeup) that you use, especially those that you use on a daily basis. There are so many natural options widely available both in store and online.

The Environmental Working Group has a great tool to search for natural options.

#4 Pay attention to toys, jewelry, and baby items

Toys and trinkets are almost always a guaranteed way to bring a smile to your child's face. Nothing wrong with that; it has always been that way throughout history and will be sure to continue. Just a few tips to make wise choices. Avoid:

- Bath toy animals that smell like plastic (not a good idea, especially in warm water)
- Toys made before lead standards were put in place, e.g., Thomas the Train and many Fisher Price Elmo toys
- Children's jewelry from arcade machines, teen stores, etc.
- Painted discount toys
- Children's pillows, blankets, and crib items that contain chemical fire retardants

TRACKING YOUR TOXINS WITH THE CAMP TOXIN TRACKER

As I promised at the beginning of this chapter, here are detailed checklists to help you eliminate the major sources of toxins in your environment.

Take one at a time (in any order!) and focus on taking any action steps that apply to that particular toxin.

Remember, this is a process and not a race. Every step you take is progress.

TOXIN	DATE
LEAD	
MERCURY	
ALUMINUM	
CADMIUM	
FLUORIDE	
CHLORINE	
FLAME RETARDANTS	
TRICHLORETHYLENE	
ZINC OXIDE NANOPARTICLES	
TITANIUM DIOXIDE NANOPARTICLES	
POLYCHLORINATED BISPHENOLS	

HOW TO REDUCE LEAD EXPOSURE

DATE: ____________

- ☐ Purchase and use a water filter. Be sure and check that your system filters out lead.
- ☐ If your house was built before 1978, have paint tested for lead.
- ☐ Get rid of all toys manufactured before 2008 or keep favorites, but do not let young children play with them.
- ☐ If you notice peeling paint on a school or neighborhood playground, say something. In the meantime, don't let your kids play on it. Old playground equipment, such as swing sets, may have several layers of paint that can peel back to expose the original lead paint. Studies have shown that soil underneath play equipment may be contaminated with lead. Older school buildings may have lead paint or lead pipes. Make it your business to have lead levels checked in water and paint.
- ☐ Avoid purchasing toys and jewelry at discount stores or vending machines.
- ☐ Do not allow young children to play with jewelry. Also, do not let children suck on zippers or buttons.
- ☐ Keep abreast of new product recalls at https://www.cpsc.gov.

Notes ____

HOW TO REDUCE MERCURY EXPOSURE

DATE: ___________

- ☐ Double-check that you do not have any old mercury thermometers around the house.
- ☐ With regard to fish consumption, the best way to reduce mercury exposure is to adhere to the following guidelines:
  - **LOWEST LEVELS OF MERCURY—eat as desired:** Catfish, clams, crab, flounder, haddock, oyster, pollock, wild salmon, sardine, scallop, shrimp
  - **MODERATE—eat maximum six times per month:** Striped bass, carp, cod, lobster, mahi mahi, perch, skate, snapper
  - **HIGHER MERCURY—eat maximum three times per month:** Croaker, halibut, Spanish mackerel, perch, sable fish, Chilean sea bass, albacore/ yellowfin tuna
  - **HIGHEST MERCURY—avoid completely:** Bluefish, grouper, king mackerel, marlin, orange roughy, swordfish
- ☐ Watch out for tuna sushi. Sushi grade tuna comes from very large tuna that contain high levels of mercury. One five-piece roll contains more mercury than you should have for over a week. Kids are especially vulnerable.

Notes ___________

HOW TO REDUCE ALUMINUM EXPOSURE

DATE: ____________

- ❑ Avoid baking powder that contains aluminum phosphate, e.g., Calumet Baking Powder or Clabber Girl Baking Powder. Instead, look for aluminum-free options like Argo, Rumford, and Whole Foods brand.
- ❑ Avoid processed foods that contain aluminum phosphate baking powder.
- ❑ Avoid foods that contain the additive sodium aluminum phosphate used in many processed cheeses, such as American cheese.
- ❑ Avoid foods that contain the additive aluminosilicate, often used as a caking agent in soup mixes and in nondairy creamers such as Coffee Mate.
- ❑ Avoid antacids that contain the ingredient aluminum hydroxide, like Maalox, Mylanta, Gaviscon, Di-Gel, Acid Gone, Gelusil.
- ❑ Avoid cooking with aluminum cookware and foil.
- ❑ Do not shave before applying aluminum-containing antiperspirant as this could result in absorption of aluminum into the bloodstream. Do not use spray-on antiperspirant, as aluminum can be inhaled into lungs.
- ❑ Avoid makeup that contains aluminum powder. Aluminum powder is especially troubling, as it consists of nanoparticles that are easily absorbed by your skin and enter your bloodstream.

Notes ____________

HOW TO REDUCE FLUORIDE EXPOSURE

DATE: ___________

- ☐ Find out if the water in your area is fluoridated. If it is, use a reverse osmosis water filter in your kitchen. Charcoal filters do not remove fluoride.
    - Dentists now recommend children only use a pea-sized amount of toothpaste rather than a strip covering the length of the brush. Supervise brushing for younger children. Make sure toothpaste is rinsed out and not swallowed.
- ☐ Avoid nonorganic foods. Due to its toxicity, fluoride is used in some pesticides to kill insects and other pests. As a result, some food products, particularly grape products, dried fruit, dried beans, cocoa powder, and walnuts have high levels of fluoride.
- ☐ Reduce intake of foods made with mechanically separated meat (e.g., chicken fingers, nuggets, etc.), especially in young children.
- ☐ Avoid nonstick pans. Cooking food or boiling water in nonstick pans may increase the fluoride content of food. One study found that boiling water in a nonstick pan for just fifteen minutes more than doubled the amount of measurable fluoride.[344]
- ☐ For adults, nonfluoride toothpaste is an option. Check out the Environmental Working Group's toothpaste database for numerous chemical-free options.

Notes ___________

____

____

____

344 Shannon IL. Effect of aluminum and teflon cooking vessels on fluoride content of boiling water. J Mo Dent Assoc. 1977 Feb;57(2):26-28. PMID: 274566.

HOW TO REDUCE CADMIUM EXPOSURE

DATE: ___________

- ☐ Avoid buying young children jewelry. The greatest potential for exposure comes from swallowing a jewelry piece. However, exposure also occurs from repeatedly biting, sucking, mouthing, and frequent handling the jewelry piece.
- ☐ Keep adult jewelry out of reach of children.
- ☐ Warn teenagers not to put costume jewelry in their mouth. Costume jewelry from chains such as ALDOs have tested positive for excessive levels of cadmium.
- ☐ Stop smoking. Cigarette smoke contains cadmium, which can be absorbed through the lungs.
- ☐ Avoid secondhand smoke

Notes ___________

HOW TO REDUCE EXPOSURE TO FIRE RETARDANTS

DATE: ____________

- ❑ Avoid contact with decaying or crumbling foam that might contain fire retardants. This includes older vehicle seats, upholstered furniture, foam mattress pads, carpet padding, and children's products made of foam. Always keep foam enclosed.
- ❑ Check labels of baby products such as nursing pillows, portable cribs, and baby carriers. Avoid polyurethane foam and seek alternatives containing cotton, wool, or polyester.
- ❑ Be careful when removing or replacing old carpet, since PBDEs are found in the foam padding beneath carpets. Have a HEPA vacuum handy.
- ❑ Fire retardants are also found in electronics. Prevent young children from touching and mouthing items with fire retardant, especially your cell phone or remote control.
- ❑ When it's time to replace furniture and mattresses, consider buying wooden furniture or furniture filled with polyester, down, wool, or cotton, as they are unlikely to contain added fire-retardant chemicals.

Notes ____________________

HOW TO REDUCE EXPOSURE TO CHLORINE

DATE: ____________

- ❑ Avoid bleach safety wipes, especially around children.
- ❑ Do not use bleach to clean bathrooms.
- ❑ Install a shower water filter if possible.
- ❑ Minimize time in indoor chlorinated pools.

Notes ____

HOW TO REDUCE EXPOSURE TO TITANIUM DIOXIDE

DATE: ____________

- ❑ Check food labels for titanium dioxide. Remember, 35% of all foods/medication/candies that list titanium dioxide as an ingredient contain nanoparticles of titanium dioxide.
- ❑ Strategically avoid the following food types when possible: M&M candy, Trident gum, Mentos candy, powdered sugar donuts, etc. Kid's foods are more likely to contain titanium dioxide nanoparticles.
- ❑ Check labels on sunscreen and cosmetics for titanium dioxide. Only buy products that say "non-nano."

Notes ____

HOW TO REDUCE EXPOSURE TO ZINC OXIDE

DATE: ____________

- ❑ Avoid sunscreen and cosmetics that have zinc oxide nanoparticles, especially application to the lips and mouth area.
- ❑ Avoid spray sunscreen.

Notes ____________

HOW TO REDUCE EXPOSURE TO TRICHLORETHYLENE

DATE: ____________

- ❑ Check with your dry cleaner to see if this type of spot cleaner is being used on your clothes. If so, switch to a dry cleaner that does not use this substance.
- ❑ Avoid arts and craft sprays, paints, paint thinners, and metal degreasers that contain trichloroethylene.
- ❑ For those living near military bases or near hazardous waste dump sites, avoid allowing kids to play in outside dirt.

Notes ____

HOW TO REDUCE EXPOSURE TO POLYCHLORINATED BISPHENOLS (PCBS)

DATE: ____________

- [ ] The greatest exposure to PCBs is through consumption of contaminated seafood. Before cooking, remove the skin and fat where toxins are likely to accumulate; when cooking, let fat drain away.

Notes ____________________

CHAPTER 20

CAMP FLEXIBLE FOOD AND NUTRITION PLAN

PRINCIPLE ONE	Eat Foods that Contain Compounds to Support Calcium ATPase
PRINCIPLE TWO	Manage Your Blood Sugar
PRINCIPLE THREE	Embrace Meal Planning and Cooking

What and how we eat is our life force. In other words, we are what we eat. While I appreciate that food access is an issue for many people, whereby quantity may be the priority rather than quality, please consider what I have to offer as my ideal. Everyone has a different starting point, and fully honoring that is important to me. But I would be remiss if I didn't push you to do better! My approach to food considers human and environmental health as a top priority. So please consider the following:

- Consume fresh whole foods, organic if possible
- Consume a plant-centric diet; if you eat animal foods, high welfare is essential

- Embrace cooking; it's an important part of connecting to nourishment and healing

Choosing food is confusing. How could it not be? Big food companies spend billions of dollars per year marketing ultraprocessed foods to consumers boasting some form of edible health when in fact, if you read between the lines, there is not. With the rise of social media, anyone can be a "health expert" these days. And if you have a strong social media following, the media pick you up and your platform gets bigger. It doesn't matter if you are credible or not. Then there are diet fads, on trend as fast as they are off trend. Many are written by health professionals, and many are written by those social media celebrities. As for me, this book started out as something for the scientific community so I could share my years of research and be an advocate for a healthy lifestyle. As I wrote, it turned more into a "how-to" for the reader, but I have kept it pretty basic, as I am neither a formally trained health professional nor an expert. Though I do have the great honor and pleasure of knowing and working with many! Simply, my end goal is for you to live and eat healthier so you can support Calcium ATPase and all of the other significant and critical nutrient needs of your human body.

All of that aside, let's be real: eating is an emotional act for many. It can be tied to good experiences and bad. This is just the reality of being human. And sometimes it's those emotions, coupled with inevitable confusion about what to choose, that get in the way of doing what is best for your health. The sum of my experiences made me question, many years ago, what I ate and what I was feeding my son. My determination to make him well put me on a path that forced me to do things differently and to feel really uncomfortable at first. Change is hard, I know. So yes, asking you to get a little uncomfortable is not taken lightly over here. But with practice, a new and healthier normal could be birthed. Please be kind and compassionate to yourself as you journey through this chapter.

For the CAMP Food and Nutrition Plan, I could have easily referred you to the many books boasting healthy meal plans and recipes from well-known medical and health professionals. However, instead I am giving you a very special opportunity to learn from one of the best when it comes to food choice and health outcome: culinary nutritionist and author of *What The Fork Are You Eating* (© 2014 TarcherPerigee | Penguin Random House) Stefanie Sacks, MS, CNS, CDN. She has been studying food and healing for over three decades; is a professionally trained chef mentored by the great Annemarie Colbin, PhD, and the Natural Gourmet Institute for Health and Culinary Arts (now part of Institute for Culinary Education); and has her Master of Science in Nutrition Education from Teachers College, Columbia University. She is also a Certified Nutrition Specialist and Certified Dietitian Nutritionist. As I have come to learn, Stefanie is someone who lives with illness and turned to food in the 1980s to make a difference in her health. In other words, she walks the walk and talks the talk. And was doing it way before it was hip and cool. Also, her younger son had numerous health challenges from birth onward like my son, Knute, and she tirelessly fought for his health and turned to food to make a difference. If you stick with me through Principles One and Two, Stefanie will take you through Principle Three ending with ten delicious and nutritious Calcium ATPase Supportive Recipes. This is just a start, so I strongly encourage you to visit Stefanie's website at stefaniesacks.com and to purchase her book (also available in audio), as our approach to food and its role in health is simpatico.

PRINCIPLE ONE

Eat Foods that Contain Compounds to Support Calcium ATPase

The good news is that Mother Nature has given us lots of options to support Calcium ATPase!

To stay consistent, I will start with the research. Bear with me, and it will lead you to the translation to the plate and, further on, Stefanie's passion piece. It's important to give you the science behind my suggestions so that you can believe (maybe this is a projection; I am a natural skeptic!) that this is not just a list of generally healthy foods (which it is, as well!). The bottom line is that in addition to supporting Calcium ATPase, research shows that the majority of these compounds have a number of other positive health effects. I don't want to put you to sleep with all the good they can do, so I am just sticking to their impact on

Calcium ATPase. Each compound highlighted below begins with the science, then takes you to the plate. Hopefully, this finer accounting of where you can find Calcium ATPase supportive compounds in common foods will make this process of shifting what and how you eat a little more digestible.

VITAMIN E

The Science

Vitamin E has positive effects on Calcium ATPase in the brain, liver, kidney, skeletal muscles, and heart, including:

- Prevents a decline in Calcium ATPase activity in high cholesterol environments, including the brain, heart, kidneys, and skeletal muscles.[345]
- Maintains Calcium ATPase levels in the heart during stress, excessive alcohol consumption, and following a heart attack.[346,347,348]
- Helps prevent the decline in brain Calcium ATPase levels that can occur with diabetes.[349]
- Prevents damage to Calcium ATPase during toxin exposure, such as from organophosphate insecticides.[350]

345 Ademoglu E, Gokkusu C, Palanduz S. Vitamin E and ATPases: protection of ATPase activities by vitamin E supplementation in various tissues of hypercholesterolemic rats. Int J Vitam Nutr Res 2000 Jan; 70(1):3–7.
346 Nanji A, Sadrzadeh S. Effect of fish oil and vitamin E on ethanol-induced changes in membrane ATPases. Life Sc. 1994;55(12):PL245–49.
347 Reddy VD, Padmavathi P, Bulle S, Hebbani AV, Marthadu SB, Venugopalacharyulu NC, Maturu P, Varadacharyulu NC. Association between alcohol-induced oxidative stress and membrane properties in synaptosomes: A protective role of vitamin E. Neurotoxicol Teratol. 2017 Sep;63:60–65. doi: 10.1016/j.ntt.2017.07.004. Epub 2017 Aug 1. PMID: 28778836.
348 Qin F, Yan C, Patel R, Liu W, Dong E. Vitamins C and E attenuate apoptosis, beta-adrenergic receptor desensitization, and sarcoplasmic reticular Ca2+ ATPase downregulation after myocardial infarction. Free Radic Biol Med. 2006 May 15;40(10):1827–42. doi: 10.1016/j.freeradbiomed.2006.01.019. Epub 2006 Feb 8. PMID: 16678021.
349 Das Evcimen N, Ulusu N, Karasu C, Dogru B. Adenosine triphosphatase activity of streptozotocin-induced diabetic rat brain microsomes. Effect of vitamin E. Gen Physiol Biophys. 2004 Sep;23(3):347–55.
350 Bhatti G, Bhatti J, Kiran R, Sandhir R. Alterations in Ca2+ homeostasis and oxidative damage induced by ethaion in erythrocytes of Wistar rats: ameliorative effects of vitamin E. Environ Toxicol Pharmacol 2011 May; 31(3):378–86.

The Action: To the Plate

These foods are either rich in Vitamin E or contain quantities worth mentioning. You will see that many of these ingredients are included in the recipes Stefanie created. See 3-Quick Vitamin E Snacks below.

FOOD SOURCES		
Almonds	Olive Oil	Sunflower Seeds
Avocado	Palm Oil	Sweet Potato
Butternut Squash	Peanuts, Peanut Butter	Tomatoes
Hazelnuts	Pine Nuts	Turnip Greens
Mango	Spinach	Wheat Germ Oil

3-QUICK VITAMIN E-RICH SNACKS
Trail mix with sunflower seeds, almonds, hazelnuts and peanuts; you can toss in some dark chocolate or raisins for a little sweet
Mashed avocado (with olive oil and salt) on crackers or toast
Baby tomatoes

ELLAGIC ACID

The Science

Ellagic acid is a natural antioxidant found primarily in fruits and nuts. It has a positive impact on Calcium ATPase in the following ways:

- Ellagic acid stimulates Calcium ATPase in the diabetic heart.[351]
- Ellagic acid stimulates Calcium ATPase in the nondiabetic heart.[352, 353]
- Ellagic acid normalizes Calcium ATPase levels in the kidneys after exposure to chemical toxins.[354]

The Action: To the Plate

These foods are either rich in ellagic acid or contain quantities worth mentioning. You will see that many of these ingredients are included in the recipes Stefanie created. See 3-Quick Ellagic Acid Snacks below.

FOOD SOURCES	
Blackberries	Raspberries
Chestnuts	Red grapes
Cranberries	Red wine
Green tea	Strawberries
Pecans	Walnuts
Pomegranate juice	

351 Namekata I, Hamaguchi S, Wakasugi Y, Ohhara M, Hirota Y, Tanaka H. Ellagic acid and gingerol, activators of the sarco-endoplasmic reticulum Ca^{2+}-ATPase, ameliorate diabetes mellitus-induced diastolic dysfunction in isolated murine ventricular myocardia. Eur J Pharmacol. 2013 Apr 15;706(1-3):48–55. doi: 10.1016/j.ejphar.2013.02.045. Epub 2013 Mar 13. PMID: 23499698.

352 Antipenko AY, Spielman AI, Kirchberger MA. Interactions of 6-gingerol and ellagic acid with the cardiac sarcoplasmic reticulum Ca2+-ATPase. J Pharmacol Exp Ther. 1999 Jul;290(1):227–34. PMID: 10381780.

353 Berrebi-Bertrand I, Lahouratate P, Lahouratate V, Camelin JC, Guibert J, Bril A. Mechanism of action of sarcoplasmic reticulum calcium-uptake activators--discrimination between sarco(endo)plasmic reticulum Ca2+ ATPase and phospholamban interaction. Eur J Biochem. 1997 Aug 1;247(3):801–09. doi: 10.1111/j.1432-1033.1997.t01-1-00801.x. PMID: 9288900.

354 Vijaya Padma V, Kalai Selvi P, Sravani S. Protective effect of ellagic acid against TCDD-induced renal oxidative stress: modulation of CYP1A1 activity and antioxidant defense mechanisms. Mol Biol Rep. 2014 Jul;41(7):4223–32. doi: 10.1007/s11033-014-3292-5. Epub 2014 Feb 25. PMID: 24566691.

3-QUICK ELLAGIC ACID SNACKS
Bowl of berries (raspberries, blackberries, and strawberries)
Trail mix with pecans, walnuts, and dried cranberries
Cup of green tea

LUTEOLIN

The Science

Luteolin is a bioflavonoid found primarily in vegetables and herbs. Luteolin has a positive impact on Calcium ATPase in the following ways:

- Luteolin increases levels of Calcium ATPase during heart failure.[355]
- Luteolin increases Calcium ATPase activity post-heart attack improving cardiac function.[356,357]

The Action: To the Plate

These foods are either rich in luteolin or contain quantities worth mentioning. You will see that many of these ingredients are included in the recipes Stefanie created. See 3-Quick Luteolin Snacks on the next page.

355 Nai C, Xuan H, Zhang Y, Shen M, Xu T, Pan D, Zhang C, Zhang Y, Li D. Luteolin Exerts Cardioprotective Effects through Improving Sarcoplasmic Reticulum Ca(2+)-ATPase Activity in Rats during Ischemia/Reperfusion In Vivo. Evid Based Complement Alternat Med. 2015;2015:365854.
356 Zhu S, Xu T, Luo Y, Zhang Y, Xuan H, Ma Y, Pan D, Li D, Zhu H. Luteolin Enhances Sarcoplasmic Reticulum Ca2+ATPase Activity through P38 MAPK Signaling thus Improving Rat Cardiac Function after Ischemia/Reperfusion. Cell Physiol Biochem 2017;41(3):999–1010.
357 Du Y, Liu P, Xu T, Pan D, Zhu H, Zhai N, Zhang Y, Li D. Luteolin Modulates SERCA2a Leading to Attenuation of Myocardial Ischemia/ Reperfusion Injury via Sumoylation at Lysine 585 in Mice. Cell Physiol Biochem. 2018;45(3):883–98. doi: 10.1159/000487283. Epub 2018 Feb 2. PMID: 29421780.

FOOD SOURCES	
Artichoke	Radicchio
Broccoli	Red leaf lettuce
Celery	Rosemary
Chicory greens	Spinach
Chili green peppers	Green bell peppers
Parsley	Yellow bell peppers
Pumpkin	Thyme

3-QUICK LUTEOLIN SNACKS
Raw broccoli dipped in your favorite dressing
Celery sticks (I love to eat them with raw almond butter!)
Bowl of green and yellow peppers

LYCOPENE

The Science

Lycopene is a carotenoid found in brightly colored fruits and vegetables. It has a positive effect on Calcium ATPase in the following ways:

- Lycopene protects brain cells from the damaging effects of environmental toxins. For example, exposure of mice hippocampal neurons to Cadmium results in a 50% reduction in Calcium ATPase. Pre-treatment with lycopene prior to cadmium exposure prevented the decline in Calcium ATPase thus promoting cell viability.[358]
- Lycopene protects cardiac and liver cells from the negative effects of atrazine (a pesticide) on Calcium ATPase.[359,360]

The Action: To the Plate

These foods are either rich in lycopene or contain quantities worth mentioning. You will see that many of these ingredients are included in the recipes Stefanie created. See 3-Quick Lycopene Snacks below.

FOOD SOURCES	
Asparagus	Red cabbage
Goji berry	Red bell peppers
Papaya	Tomatoes
Pink grapefruit	Watermelon

358 Zhang F, Xing S, Li Z. Antagonistic effects of lycopene on cadmium-induced hippocampal dysfunctions in autophagy, calcium homeostatis and redox. Oncotarget. 2017 Jul 4;8(27):44720–31. doi: 10.18632/oncotarget.18249. PMID: 28615536; PMCID: PMC5546513.
359 Lin J, Li HX, Xia J, Li XN, Jiang XQ, Zhu SY, Ge J, Li JL. The chemopreventive potential of lycopene against atrazine-induced cardiotoxicity: modulation of ionic homeostasis. Sci Rep. 2016 Apr 26;6:24855. doi: 10.1038/srep24855. PMID: 27112537; PMCID: PMC4845055.
360 Lin J, Zhao HS, Xiang LR, Xia J, Wang LL, Li XN, Li JL, Zhang Y. Lycopene protects against atrazine-induced hepatic ionic homeostasis disturbance by modulating ion-transporting ATPases. J Nutr Biochem. 2016 Jan;27:249–56. doi: 10.1016/j.jnutbio.2015.09.009. Epub 2015 Sep 25. PMID: 26476475.

3-QUICK LYCOPENE SNACKS
Small bowl fresh papaya
Sliced grapefruit
Slice of watermelon

RESVERATROL

The Science

Resveratrol, a compound produced by plants, has a positive impact on Calcium ATPase in the following ways:

- Protects the heart Calcium ATPase levels during cardiac trauma.[361]
- Stimulates Calcium ATPase activity in the hearts of people with diabetes.[362]
- Protects Calcium ATPase levels in the heart from bacterial endotoxins' harmful effects related to sepsis, a severe and sometimes fatal condition.[363]
- Counteracts some of the adverse effects pancreatitis has on Calcium ATPase in the pancreas and lungs.[364]
- Maintains spinal cord Calcium ATPase levels after a spinal cord injury and supports recovery.[365]

361 Shen M, Wu R, Zhao L, Li J, Guo H, Fan R, Cui Y, Wang Y, Yue S, Pei J. Resveratrol attenuates ischemia/reperfusion injury in neonatal cardiomyocytes and its underlying mechanism. PloS One 2012; 7(12):e51223.
362 Sulaiman M, Matta M, Sunderesan N, Gupta M, Periasamy M, Gupta M. Resveratrol, an activator of SIRT1, upregulates sarcoplasmic calcium ATPase and improves cardiac function in diabetic cardiomyopathy. Am J Physiol Heart Circ Physiol 2010 Mar; 298(3):H833-43.
363 Bai T, Hu X, Zheng Y, Wang S, Kong J, Cai L. Resveratrol protects against lipopolysaccharide-induced cardiac dysfunction by enhancing SERCA2a activity through promoting the phospholamban oligomerization. Am J Physiol Heart Circ Physiol 2016 Oct 1; 311(4).
364 Wang L, Ma Q, Chen X, Sha H, Ma Z. Effects of resveratrol on calcium regulation in rats with severe acute pancreatitis. Eur J Pharmacol 2008 Feb 2; 580(1-2):271–76.
365 Yang Y, Piao Y. Effects of resveratrol on Ca2+, Mg(2+)-ATPase activities after spinal cord trauma in rats. Zhong Yao Cai 2002 Dec; 25(120:882–85.

The Action: To the Plate

These foods are either rich in resveratrol or contain quantities worth mentioning. You will see that many of these ingredients are included in the recipes Stefanie created. See 3-Quick Resveratrol Snacks below.

FOOD SOURCES	
Cocoa powder/dark chocolate	Red grapes
Peanuts/peanut butter	Red wine
Pistachio	Strawberries
Red grape juice	

3-QUICK RESVERATROL SNACKS
1–2 squares dark chocolate
Bowl of pistachios
Handful of grapes

If you just start with these suggested Calcium ATPase-supportive snacks, you are taking a positive step toward improving your health!

ON SUPPLEMENTS TO SUPPORT CALCIUM ATPASE LEVELS

While I am a firm believer that food should always be your first line of defense, there are a few compounds whereby it's difficult to get a therapeutic dosage simply through diet. Please be sure to defer to your healthcare team when it comes to any type of supplementation. They are essentially unregulated "drugs" even though seemingly "natural," and if not sourced and administered with professional guidance, they could be harming versus doing the good intended.

What I am offering is a suggestion based on years of tireless research and my own experiences. Let this be a good place to start a conversation with your team, and take it from there.

Alpha-Lipoic Acid

Alpha-lipoic acid is a powerful antioxidant found in the mitochondria of all human cells. It helps turn nutrients into energy. Alpha-lipoic acid supports Calcium ATPase levels in the following ways:

- Protects Calcium ATPase activity in red blood cells during hyperglycemia, which protects the kidneys and arteries from the damaging effects.[366,367]
- Protects Calcium ATPase levels in the brain from toxins such as mercury and lead. [368,369,370]
- Maintains kidney Calcium ATPase activity during chemotherapy treatment with Adriamycin.[371]

366 Thirunavukkarasu V, Anitha Nandhini AT, Anuradha CV. Lipoic acid attenuates hypertension and improves insulin sensitivity, kallikrein activity and nitrite levels in high fructose-fed rats. J Comp Physiol B. 2004 Nov;174(8):587–92. doi: 10.1007/s00360-004-0447-z. Epub 2004 Sep 29. PMID: 15565449.

367 Jain SK, Lim G. Lipoic acid decreases lipid peroxidation and protein glycosylation and increases (Na(+) + K(+))- and Ca(++)-ATPase activities in high glucose-treated human erythrocytes. Free Radic Biol Med. 2000 Dec;29(11):1122–28. doi: 10.1016/s0891-5849(00)00410-x. PMID: 11121719.

368 Yang T, Xu Z, Liu W, Feng S, Li H, Guo M, Deng Y, Xu B. Alpha-lipoic acid reduces methylmercury-induced neuronal injury in rat cerebral cortex via antioxidation pathways. Environ Toxicol. 2017 Mar;32(3):931–43. doi: 10.1002/tox.22294. Epub 2016 Jun 14. PMID: 27298056.

369 Yang T, Xu Z, Liu W, Xu B, Deng Y. Protective effects of Alpha-lipoic acid on MeHg-induced oxidative damage and intracellular Ca(2+) dyshomeostasis in primary cultured neurons. Free Radic Res. 2016;50(5):542-56. doi: 10.3109/10715762.2016.1152362. Epub 2016 Mar 17. PMID: 26986620.

370 Sivaprasad R, Nagaraj M, Varalakshmi P. Combined efficacies of lipoic acid and meso-2,3-dimercaptosuccinic acid on lead-induced erythrocyte membrane lipid peroxidation and antioxidant status in rats. Hum Exp Toxicol. 2003 Apr;22(4):183-92. doi: 10.1191/0960327103ht335oa. PMID: 12755469.

371 Malarkodi KP, Balachandar AV, Varalakshmi P. The influence of lipoic acid on adriamycin induced nephrotoxicity in rats. Mol Cell Biochem. 2003 May;247(1-2):15–22. doi: 10.1023/a:1024118519596. PMID: 12841626.

Animal products, including organ and red meat, are good alpha-lipoic acid sources, but plant foods like spinach, broccoli, tomatoes, and brussels sprouts also provide it. If you don't think you're getting enough from your diet, consult with your healthcare team and consider a dosage of 600 to 1800 mg daily.

Taurine

Taurine is essential for health. It's an amino acid abundant in muscle, the brain, and many other tissues in the body. Taurine supports Calcium ATPase in the following ways:

- Maintains red blood cell calcium levels in high glucose states, or hyperglycemia.[372]
- Maintains Calcium ATPase activity in the heart during heart failure.[373] Reduced taurine is associated with reduced Calcium ATPase activity in the heart.[374,375]
- Maintains Calcium ATPase when someone has high levels of an amino acid called homocysteine often associated with heart disease.[376]
- Protects Calcium ATPase activity in the heart from chemotherapy's harmful effects.[377]

Taurine is found in animal products like fish, seafood, meat, poultry, eggs, and dairy products. However, research has shown that taurine supplementation can have some profoundly positive health benefits in addition to its role in Calcium ATPase metabolism. Supplementation with 500mg–1000 mg daily is the standard recommendation. This is one supplement that I take and strongly believe in. But again, check with your healthcare team.

372 Nandhini T, Anuradha C. Inhibition of lipid peroxidation, protein glycation, and elevation of membrane ion pump activity by taurine in RBC exposed to high glucose. Clin Chim Acta 2003 Oct: 336(1-2):129–35.
373 Zhang X, Qu Y, Zhang T, Zhang Q. Antagonism for different doses of taurine on calcium overload in myocardial cells of diastole heart failure rat model. Zhongguo Zhong Yao Za Zhi 2009 Feb; 34(3):328–31.
374 Harada H, Allo S, Viyuoh N, Azuma J, Takahashi K, Schaffer S. Regulation of calcium transport in drug-induced taurine-depleted hearts. Biochim Biophys Acta 1988 Oct 6; 944(2):273–78.
375 Ramila K, Jong C, Pastukh V, Ito T, Azuma J, Schaffer S. Role of protein phosphyorylation in excitation-contraction coupling in taurine deficient hearts.. Am J Physiol Heart Circ Physiol 2015 Feb 1; 308(3):H232-39.
376 Chang L, Xu J, Yu F, Zhao J, Tang X, Tang C. Taurine protected myocardial mitochondria injury induced by hyperhomocysteinemia in rats. Amino Acids 2004 Aug; 27(1):37–48.
377 Huang X, Zhu W, Kang M. Study on the effect of doxorubicin on expression of genes encoding myocardial sarcoplasmic reticulum Ca2+ transport proteins and the effect of taurine on myocardial protection of rabbits. J Zhejiang Univ Sci 2003 Jan-Feb; 4(1):114–20.

Green Tea

Studies have shown that a compound in green tea, epigallocatechin gallate (EGCG), supports Calcium ATPase in the following ways:

- Helps to maintain Calcium ATPase levels in the normal and diabetic heart.[378,379]
- Also maintains platelet Calcium ATPase levels, which inhibit dangerous blood clotting.[380,381]
- Protects Calcium ATPase activity from drug-induced kidney damage.[382]
- Increases Calcium ATPase levels after period of exhaustive exercise.[383]

You can glean the benefit of green by drinking green tea daily. A single cup (8 ounces or 250 ml) of brewed green tea typically contains about 50–100 mg of EGCG. Dosages used in scientific studies are often much higher, though, so if you are not a big tea drinker, consider a green tea extract with EGCG. Recommended dosage is around 300mg of EGCG daily if taking a supplement. Please check with your healthcare provider before adding this as a supplement.

Ginger

Ginger can be used in many dishes such as stir fries, curries, soups, and muffins. Ginger shots can be added to smoothies and ginger tea. The advantage of a gingerol supplement is that it can be standardized. Many nutrition experts suggest taking 200 mg daily

378 Bocchi L, Savi M, Naponelli V, Vilella R, Sgarbi G, Baracca A, Solaini G, Bettuzzi S, Rizzi F, Stilli D. Long-Term Oral Administration of Theaphenon-E Improves Cardiomyocyte Mechanics and Calcium Dynamics by Affecting Phospholamban Phosphorylation and ATP Production. Cell Physiol Biochem. 2018;47(3):1230–43. doi: 10.1159/000490219. Epub 2018 Jun 15. PMID: 29913456.

379 Babu P, Sabitha K, Shyamaladevi C. Green tea impedes dyslipidemia, lipid peroxidation, protein glycation and ameliorates Ca2+-ATPase and Na+/K+-ATPase activity in the heart of streptozotocin-diabetic rats. Chem Biol Interact 2006 Aug 25; 162(2);157–64.

380 Jin YR, Im JH, Park ES, Cho MR, Han XH, Lee JJ, Lim Y, Kim TJ, Yun YP. Antiplatelet activity of epigallocatechin gallate is mediated by the inhibitionof PLCgamma2 phosphorylation, elevation of PGD2 production, and maintaining calcium-ATPase activity. J Cardiovasc Pharmacol. 2008 Jan; 51(1):45–54.

381 Jin YR, Im JH, Park ES, Cho MR, Han XH, Lee JJ, Lim Y, Kim TJ, Yun YP. Antiplatelet activity of epigallocatechin gallate is mediated by the inhibition of PLCgamma2 phosphorylation, elevation of PGD2 production, and maintaining calcium-ATPase activity. J Cardiovasc Pharmacol. 2008 Jan;51(1):45–54. doi: 10.1097/FJC.0b013e31815ab4b6. PMID: 18209568.

382 Upaganlawar A, Farswan M, Rathod S, Balaraman R. Modification of biochemical parameters of gentamicin nephrotoxicity by coenzyme Q10 and green tea in rats.Indian J Exp Biol. 2006 May; 44(5):416–18.

383 Korf EA, Kubasov IV, Vonsky MS, Novozhilov AV, Runov AL, Kurchakova EV, Matrosova EV, Tavrovskaya TV, Goncharov NV. Green Tea Extract Increases the Expression of Genes Responsible for Regulation of Calcium Balance in Rat Slow-Twitch Muscles under Conditions of Exhausting Exercise. Bull Exp Biol Med. 2017 Nov;164(1):6-9. doi: 10.1007/s10517-017-3913-9. Epub 2017 Nov 9. PMID: 29119399.

of a product standardized to 5 percent gingerol, but again, consult your healthcare team.

PRINCIPLE TWO
Manage Your Blood Sugar

I am certainly not the first person to tell you this, and I am surely not the last. Emphasizing that high blood sugar is toxic to Calcium ATPase is critical. At the end of the day, there are several research-backed approaches to blood sugar control that include Keto (the most direct, but a plan that really needs supervision from a qualified health professional if on medication for blood sugar control), Paleo, and the Mediterranean diet. If you are currently following one of these and it's working for you, cheers! Just be sure to include Calcium ATPase supportive foods (noted above in "To the Plate") and follow the rest of the CAMP Flexible Food and Nutrition Plan.

Following, under Principle Three, you will see the CAMP 7-Day Meal & Snack Map. In its entirety, it is designed to help you assuage blood sugar highs and lows using simple food combining as its foundation. In other words, while carbohydrate counting is not integrated, it's mapped out to marry the carbs with protein and fat to ensure blood sugar balance. But to truly understand if your blood sugar is effectively being managed, consider having your A1C levels checked every couple of months. You want to aim to have your levels at or below 5.6%. If you have metabolic syndrome or full-blown diabetes and are taking medication to control your blood sugar, it's very important that you monitor it closely when making dietary changes. You may need to change your dosing. But it is critical that you always do this under the supervision of your doctor, and I can't emphasize that enough!

Oftentimes blood sugar imbalance goes hand in hand with overweight and obesity. Both result in lower Calcium ATPase levels, and lower Calcium ATPase levels reduce your metabolism. Admittedly, I am neither overweight nor obese, but having had an eating disorder in my teens, and subsequent dieting for years, I know it's a complex issue and one that I am not going to take on in this book. However, it would be irresponsible of me to avoid mentioning the link between reduced Calcium ATPase and weight. You are more susceptible to diseases such as diabetes and heart disease if you are overweight. This is not new news! However, this is *not* a diet book or a weight loss plan. My focus is educating you about Calcium ATPase and its critical role in our health, then giving you tools to support healthy production in your body. It's really about making lifestyle changes!

If you need to lose significant weight, use this book as a guide to jump-start some new (or maybe not so new) ideas. Let it serve to inspire you to begin to make small changes. The CAMP 7-Day Meal & Snack Map is a great place to begin! And if you follow, you could very well lose some weight while you, most importantly, gain health.

PRINCIPLE THREE

Embrace Meal Planning and Cooking

by Culinary Nutritionist Stefanie Sacks, MS, CNS, CDN

When Brunde approached me about creating recipes for this book, admittedly I was suspect. As a seasoned nutritionist, I have seen hundreds of edible ideals and fads jump in and out of popularity. And I don't attach myself to any of them. Ever. But instead of cutting her off at "hello," I chose to listen, ask questions, and dive a little deeper. And what I came up with is this—Brunde is a mom who tirelessly and passionately worked to help her suffering child through immense and mysterious health challenges, something I can profoundly relate to. And now she wants to share what she uncovered with the hope that it can help as many people as possible. Who can argue with that?

While it's not necessarily my shtick to put so much energy into one seemingly small enzyme, what Brunde is essentially asking of you is *exactly* what I asked of myself beginning in the 1980s and have asked of my clients and students since 1999 and my readers (and listeners) since 2014 when *What The Fork are You Eating* was released—to learn about where environmental and edible toxins lie that negatively impact health on a daily basis and to do something about them, even if small. Food and how it is grown and processed, quite unfortunately, is one of the biggest culprits. So we are both aiming to inspire and offer up an education based on science and, believe it or not, logic, which hopefully will empower you to think differently about how you choose food.

If you have made it this far, you know that Brunde is a "science geek." And believe it or not, so am I. But I am also a passionate chef who has learned through personal experiences as well as formal education how to translate the science to the plate to get healthy and stay healthy. Just as I have done for decades with so many, my aim is the same here—to give you practical tools to make a difference in your health through food choice!

I have created a series of tools including:

- CAMP Edible Resources
- CAMP 7-Day Meal & Snack Plan
- CAMP Ingredient Map
- CAMP Essential Food Shopping Guide
- CAMP Recipes

These are all intended to jump-start your Calcium ATPase supportive journey. That being said, this has been and is a true collaboration and one that I am honored to be a part of. Please know that all of these recipes were created with love specifically for this book, and I hope that you enjoy them as much as I relished developing, testing, and tasting each one.

Finally, while cooking can be very scientific, it can also be a place to let go and get creative. Recipes are guides only, so please sink into the experience of food shopping, cooking, and eating, as the entire process can be wildly nourishing. Our collective goal is to nourish!

CAMP EDIBLE RESOURCES

With these resources, I aim to give you a handful of tips on how to healthfully choose food. For more of an encyclopedic guide, you can read or listen to my book. I have picked my top five things to consider when buying food:

1. Buy organic plant foods when and if you can, paying close attention to the Dirty Dozen and Clean Fifteen by the Environmental Working Group (EWG), as Brunde outlined in the previous chapter. There is a great smartphone app that is an invaluable guide.
2. As for animal foods, choosing truly high welfare edibles can be very confusing and challenging. There are two certifications that are my go-to: Animal Welfare Approved by A Greener World and Certified Humane by Humane Farm Animal Care. In addition, knowing your farmer is a great option.
3. The ingredient lists tell the story of your food, not the Nutrition Facts, so please read the lists before you opt in and make sure you know what each ingredient is. If the list is more than an inch long and you can't pronounce an item, you shouldn't buy the product.
4. Don't be fooled by the health claims and label lingo, as they are, at best, misleading and, at worst, flat-out lies (I go into great detail about this in my book; you can also see Food Labels Exposed by A Greener World at agreenerworld.org/resources/food-labels/).
5. Opt into a whole fresh foods-based diet 70–80% of the time.

I hope these pointers can help you rethink food and be a catalyst for change!

CAMP 7-DAY MEAL & SNACK MAP

This meal and snack map should only serve as a guide to reboot your health and ensure that what you choose to eat is chock-full of Calcium ATPase supportive ingredients. It is not intended as a weight loss protocol or a detoxification plan. However, you may very well lose weight and remove toxins along the way. Not a bad thing! Adapt as you need and dive in for the short term. It can also be a map for the long term. But I urge you to stay connected to Brunde at BrundeBrody.com. This CAMP 7-Day Meal & Snack Map will grow and evolve as we continue this journey with you! A few things to note:

- Stay hydrated by drinking water throughout the day; coconut water is also a great option!
- If you are an avid coffee drinker (or not), try two cups of green tea or one cup of matcha green tea per day as an alternative; it's a great Calcium ATPase booster.
- Portions are only intended as a guide; eat as many vegetables as you please!
- This Map is carefully planned so you can integrate leftovers into your everyday. Not only does it save time cooking, but it also helps mitigate food waste (a pet peeve of mine).
- Many of the recipes can be doubled, portioned, and frozen for consumption within a 2–3-month period. For many, this is a good way to be efficient in the kitchen.
- The **bolded** items in the Map are recipes created for this book.
- If you want similar recipes that are not necessarily designated as Calcium ATPase supportive but are essentially just that, you can see fifty of them in my book, *What The Fork are You Eating.*
- The snacks offer a variety of quick picks to choose from. I suggest 2–3 snacks per day!

THE CALCIUM CONNECTION

	BREAKFAST	LUNCH	DINNER	SNACKS
MONDAY	8oz **Berry Coconut Smoothie Bowl**	Large spinach salad with ½ sliced avocado, tomatoes, celery, red onion, sunflower seeds or slivered almonds (about 2 cups packed) with lemon and extra virgin olive oil	1 packed cup **Red Leaf Slaw with Pomegranate Dressing** 1 cup **Creamed Rosemary Asparagus Soup** 2–4oz fish, poultry, meat, or vegetarian option roasted, sautéed, or grilled	Trail mix with sunflower seeds, almonds, hazelnuts, pecans, walnuts, dried cranberries Bowl of pistachios Apple with 1 tablespoon or peanut or almond butter 1 sliced grapefruit 1 cup of berries (raspberries, blackberries, and strawberries) 1 cup of fresh grapes, papaya, or watermelon Baby tomatoes Bowl of green, yellow, and red bell peppers Raw broccoli dipped in your favorite dressing Celery sticks with 1 tablespoon raw almond butter ½ Mashed avocado (with olive oil and salt) on crackers or toast 1–2 squares dark chocolate
TUESDAY	1 slice whole-grain toast with ¼ avocado mashed, topped with dash of olive oil, salt, and ground cumin	1 packed cup **Red Leaf Slaw with Pomegranate Dressing** 1 cup **Creamed Rosemary Asparagus Soup**	1 cup **Black Lentil Veggie Salad with Ginger Grapefruit Vinaigrette** 2–4oz fish, poultry, meat, or vegetarian option roasted, sautéed, or grilled	
WEDNESDAY	1 serving **Green Eggs**	1 cup **Black Lentil Veggie Salad with Ginger Grapefruit Vinaigrette** with canned salmon or sardines	Oven-roasted red cabbage* ½ cup **Sweet Potato Sundried Tomato Risotto** 2–4oz fish, poultry, meat, or vegetarian option roasted, sautéed, or grilled	
THURSDAY	8oz **Berry Coconut Smoothie Bowl**	1 cup **Creamed Rosemary Asparagus Soup** Large spinach salad with ½ sliced avocado, tomatoes, celery, red onion, sunflower seeds or slivered almonds with lemon and extra virgin olive oil (about 1 cup packed)	½ cup steamed quinoa or 1 small baked sweet potato 2–4oz **Ratatouille-Smothered Wild Salmon**	

CAMP Flexible Food and Nutrition Plan

	BREAKFAST	LUNCH	DINNER	SNACKS
FRIDAY	½ cup cooked oatmeal topped with berries and walnuts (about ¼ cup dry oats cooked with water); add cinnamon and 1 teaspoon honey or maple syrup	Greens with 2–4oz **Ratatouille-Smothered Wild Salmon**	1 cup steamed or sautéed broccoli ½ cup steamed quinoa 2–4 ounces **Drunken Slow-Roasted Chicken**	**SEE OPTIONS ABOVE**
SATURDAY	1 serving **Green Eggs**	Bed of spinach with 2–4oz **Drunken Slow-Roasted Chicken**	Bowl of baby tomatoes and cucumbers with lemon, extra virgin olive oil, and salt ½ cup steamed brown or white rice 1 cup **Butternut Squash Turkey Chili**	
SUNDAY	1 cup plain yogurt or dairy-free option topped with berries and pecans	**Butternut Squash Turkey Chili Burrito**	2–3 slices **Deep Dish Gluten Free Veggie Pizza**	

**To make oven-roasted red cabbage, slice the leftover cabbage from Red Leaf Lettuce Slaw recipe into ½-inch slabs. Place on baking sheet and drizzle with extra virgin olive oil and ½ teaspoon of salt. Roast in preheated 375° oven for 15 minutes. Flip and roast for another 15 minutes.*

CAMP INGREDIENT MAP

Each recipe below is laced with nutrient-dense ingredients that are health supportive and anti-inflammatory. I use alliums (onions, garlic, leek, scallion . . .) generously, as not only do they add remarkable flavor to a dish, but also powerful nutrition. While I don't want to micromanage nutrients, as I do believe that nutrition analysis can lead to culinary paralysis, helping you understand where Calcium ATPase supportive ingredients live is important. So I have created an ingredient map to simplify matters. Then onto shopping and cooking!

Ingredients	Vitamin E	Ellagic Acid	Luteolin	Lycopene	Resveratrol
BERRY COCONUT SMOOTHIE BOWL					
Raspberries		X			
Blackberries		X			
Strawberries		X			X
Celery			X		
Avocado	X				
Pecans		X			
GREEN EGGS					
Olive oil	X				
Spinach	X		X		
Parsley			X		
CREAMED ROSEMARY ASPARAGUS SOUP					
Asparagus				X	
Rosemary			X		
Olive oil	X				

Ingredients	Vitamin E	Ellagic Acid	Luteolin	Lycopene	Resveratrol
RED LEAF SLAW WITH POMEGRANATE DRESSING					
Red leaf lettuce			X		
Radicchio			X		
Red cabbage				X	
Artichoke hearts			X		
Grape tomatoes	X			X	
Shelled pistachios					X
Pomegranate juice		X			
Celery			X		
Avocado	X				
Parsley			X		
Olive oil	X				
BLACK LENTIL VEGGIE SALAD WITH GINGER GRAPEFRUIT VINAIGRETTE					
Olive oil	X				
Celery			X	X	
Parsley			X		
Sunflower seeds	X				
Red grapes					X
Grapefruit				X	
Ginger			X		
SWEET POTATO SUNDRIED TOMATO RISOTTO					
Olive oil	X				
Sweet potato	X				
Sundried tomatoes				X	
Parsley			X		
Red wine					X
Rosemary			X		

Ingredients	Vitamin E	Ellagic Acid	Luteolin	Lycopene	Resveratrol
DEEP-DISH GLUTEN-FREE VEGGIE PIZZA					
Olive oil	X				
Tomato sauce				X	
Broccoli			X		
Spinach			X		
Sundried tomatoes				X	
Sweet green peppers (bell)			X		
BUTTERNUT SQUASH TURKEY CHILI					
Olive oil	X				
Butternut squash	X				
Diced tomatoes				X	
Yellow peppers			X		
RATATOUILLE-SMOTHERED WILD SALMON					
Olive oil	X				
Baby tomatoes			X		
Yellow pepper			X		
Chili peppers			X		
Spinach	X				
DRUNKEN SLOW-ROASTED CHICKEN					
Olive oil	X				
Pine nuts	X				
Baby tomatoes				X	
Parsley			X		
Red wine					X
Cranberries		X			
Rosemary			X		
Thyme			X		

CAMP ESSENTIAL FOOD SHOPPING GUIDE

This is not, specifically, a shopping list for the recipes, as that may overwhelm you. However, it is enough to serve as a foundation for each recipe, and it may even renew your pantry, fridge, and freezer along the way. Plus, I tossed a few must-haves in there for you, too! For the most part, items are categorized by grocery aisle to make your jaunt to the market easier.

GROCERY AISLE: PRODUCE			
Apples	Lemons	Asparagus	Onions (leeks, scallion, chives, shallots)
Avocado	Papaya	Parsley	Baby tomatoes
Potatoes, russet	Bell peppers (red, yellow, green)	Raspberries	Blackberries
Red grapes	Broccoli	Red leaf lettuce	Butternut squash
Rosemary	Scallions	Celery	Shallots
Cranberries	Spinach	Garlic	Strawberries
Ginger	Sweet potatoes	Grape tomatoes	Thyme
Grapefruit	Watermelon		
GROCERY AISLE: REFRIGERATED & FROZEN			
Butter +/or ghee (clarified butter)	Ground turkey	Chicken breasts, boneless & on the bone	Milk or nondairy version
Eggs	Plain yogurt or nondairy version	Fresh/frozen fish, caught and processed in the US	Tofu +/or tempeh

GROCERY AISLE: GRAINS & LEGUMES			
Arborio rice	Short grain brown rice	Basmati rice, brown +/or white	Tortillas
Lentils, black (green and red good, too)	White beans, canned	Oats, rolled	Whole grain or gluten-free bread
Quinoa			

GROCERY AISLE: NUTS, NUT BUTTERS & SEEDS			
Almond butter	Pecans	Almonds	Pine nuts
Cashews, raw	Pistachios	Flax meal	Sunflower seeds
Peanut butter (no sugar added)	Walnuts	Peanuts	

GROCERY AISLE: DRIED HERBS & SPICES			
Black pepper	Red pepper, crushed	Cumin, ground	Rosemary
Garlic powder	Sea salt	Ginger powder	Thyme
Onion powder	Turmeric	Parsley	

GROCERY AISLE: CANNED & CONDIMENTS			
Artichoke hearts	Olive oil, extra virgin cold pressed	Coconut cream or full fat coconut milk	Pomegranate juice
Coconut, shredded unsweetened	Red wine	Dark chocolate	Sundried tomatoes
Diced tomatoes, canned	Tomato sauce	Green tea	Vinegar, balsamic, red, white +/or rice
Honey	Vegetable broth		

CAMP RECIPES

BERRY COCONUT SMOOTHIE BOWL **SERVES 2 (8-OUNCE PORTIONS)**

My kids love acai bowls. The frozen acai, on occasion, can be found in my freezer, but my preference is sticking with local ingredients. So, consider this recipe the "reinvented acai bowl" using frozen berries. Stick with organic berries if you can, as the non-organic are often celebrities on the Environmental Working Group's Dirty Dozen. And if you have a bunch of berries going bad, trim and freeze to avoid waste. This recipe is meant for two, so if it's just you, put the second portion in an airtight container in the fridge and enjoy the next day.

INGREDIENTS

¾ cup plain rice milk (coconut water or other unsweetened nondairy milk OK, too)
¼ cup coconut cream (about ½ of 5.4 Fl Oz can)
1 cup fresh frozen berries, no added sugar (raspberries, blackberries, strawberries)
½ cup celery, thinly sliced (about 1 large stalk)
2 tablespoons ripe avocado
2 tablespoons pecans
1 tablespoon flax meal
1 teaspoon honey

METHOD

1. Add all ingredients to blender and puree until smooth.
2. Evenly divide between two bowls and top with unsweetened shredded coconut and fresh fruit. You can also add cacao nibs and muesli.

GREEN EGGS

SERVES 2–4

Nothing beats simplicity, and combine it with delicious and nutritious, I am sold! Do your best to get your hands-on organic pasture-raised eggs with a high welfare seal of approval (Animal Welfare Approved or Certified Humane are my ideal) for this recipe, as they truly have superior nutrition to any other option. If not, buying organic will do. Or find a local farm nearby. You can eat the eggs straight up, with some whole-grain toast, or wrap them with a little salsa in a warmed tortilla.

INGREDIENTS

4 eggs, beaten
2 tablespoons extra virgin olive oil
1 cup leek, thinly sliced (about ½ large leek)
3 cups baby spinach or other dark leafy greens, packed then rough chopped
¼ cup rough chopped parsley
Pinch of salt
fresh ground pepper, to taste

METHOD

1. Prep leek, spinach, and parsley.
2. In a medium pan, heat olive oil on medium and sauté leeks until tender (about 3 minutes), then add spinach and cook until wilted (about 1–2 minutes). Top with parsley and mix well.
3. Pour beaten eggs into pan and scramble with vegetables for about 3 minutes. For well-done eggs, continue to scramble for another 3 minutes.

CREAMED ROSEMARY ASPARAGUS SOUP

SERVES 4–6

This soup is super light with a texture that is likened to cashmere. And believe it or not, it's dairy free! The yield is roughly 8 cups, and while that may seem like a lot, especially for one or two people, I urge you to make the whole batch and freeze what you don't use. It will keep nicely for up to 3 months. And take some cooking "stress" off your shoulders. You can pair this with the Red Leaf Slaw for a soup-and-salad combo or any of the recipes below for a well-balanced meal. I also like to have the soup straight up with a piece of grainy gluten-free bread for dipping!

INGREDIENTS

¼ cup extra virgin olive oil
2 cups thinly sliced yellow onion (about 1 large onion)
4 cloves garlic, rough chop
2 cups russet potato, peeled and diced (about 2 medium potatoes)
2 bunches asparagus, bottoms trimmed (up to 1.5"), cut into 2" pieces
6 cups water or light flavored broth*
2 tablespoons rosemary, minced
2 teaspoons salt
½ lemon, juiced
½ cup raw cashews
fresh ground pepper, to taste

METHOD

1. Prep onion, garlic, potato, asparagus, and rosemary.
2. Using a 4-quart pot, heat oil on medium and add onion and garlic. Cook until translucent, about 3 minutes.
3. Add potato and asparagus and mix well. Then add water or broth, rosemary, and salt. Cook, covered, for 10 minutes. Soup will come to a roaring boil.
4. Then remove cover and let cook for another 10 minutes.
5. Using a high-speed blender, pour all ingredients from pot into blender and add cashews and lemon juice. Purée until creamy. Add fresh ground pepper to taste and more salt if needed.

NOTES

*On the broth, go with a low-sodium version and only use 2 cups (plus 4 cups of water). Many boxed broths are heavily flavored, meaning they dominant the taste profile of the dish. I typically use water and sometimes add a bouillon cube (make sure it's one without harmful ingredients, including MSG).

RED LEAF SLAW WITH POMEGRANATE DRESSING

SERVES 4–6

I am not a fruit-in-my salad kind of gal, but I am truly a fan of this creation—the perfect balance of savory and sweet! The ingredient list may seem daunting, but this untraditional slaw is packed with powerful nutrition and crunch and well worth the attention. And trust me when I tell you that it will take about 10 minutes to throw together for 24 hours of nourishment (just keep the dressing on the side and toss when you are ready to nosh).

INGREDIENTS

2 cups red leaf lettuce, outer leaves removed, cleaned, dried, and cut into thick shreds
1 cup thinly shredded radicchio*
1 cup thinly shredded red cabbage**
½ cup flat leaf parsley, rough chop
1 cup grape tomatoes, halved
½ cup thinly sliced celery (about 1 large stalk)
4 canned or jarred artichoke hearts, chopped***
1 cup shelled pistachios, unsalted or salted

DRESSING
¼ cup extra virgin olive oil
¼ red wine vinegar
½ cup pomegranate juice
1 clove garlic
½ avocado

METHOD

1. Combine all vegetables in large bowl and gently mix. If you are not going to eat the slaw in one sitting, store what you will eat later in the fridge in an airtight container.
2. Add all dressing ingredients in food processor or blender and purée until creamy. Toss ½ of the dressing with the vegetables and add more dressing to taste.

NOTES

You will have leftover ingredients, ones that may stump you. So here is a little extra guidance:

*For the radicchio, you can wrap in a damp paper towel and store in the fridge for up to a week; then run with this recipe again!

**For the cabbage, that too will keep for a good week, maybe even two. Or you can opt into another creation like my oven roasted red cabbage as noted in italics at the bottom of page 201.

***For the artichoke hearts, toss them into tomato sauce and top with your favorite pasta dish. You can also add to the Ratatouille-Smothered Salmon recipe below (as an add-in).

BLACK LENTIL VEGGIE SALAD WITH GINGER GRAPEFRUIT VINAIGRETTE

SERVES 4–6

The key to the deliciousness of this dish is cooking the lentils to perfection—that means leaving them slightly tough and chewy yet soft. Just follow my lead and you will do great! This nutrition powerhouse can last for three days, each day getting better! It travels well, so it's a great go-to for office lunches and picnics. Enjoy as a stand-alone or pair with the Creamed Rosemary Asparagus soup for top-notch nourishment.

INGREDIENTS

1 cup black lentils
3 cups water
1 cup thinly sliced celery (about 2 stalks)
½ cup thinly sliced scallions (about 2 large scallions)
1 cup flat leaf parsley, destemmed, loosely packed and rough chopped
½ cup sliced red seedless grapes (about 10 large grapes)
¼ cup shelled sunflower seeds

DRESSING

¼ cup extra virgin olive oil
1 tablespoon plain rice vinegar
½ cup grapefruit juice, fresh squeezed
1 teaspoon ginger root, minced
½ teaspoon salt

METHOD

1. Combine lentils and water in medium pot. Partially cover and bring to a boil on medium; reduce to simmer and cook for 15 minutes, until lentils are tough but tender.
2. Prep all vegetables and add to large bowl.
3. Combine all dressing ingredients in a small jar or container, secure lid and shake to mix.
4. When lentils are finished, pour contents of pot through fine mesh strainer and run under cold water. Once cooled, combine with vegetables, dressing, and toss well.

SWEET POTATO SUNDRIED TOMATO RISOTTO

SERVES 4–6

I am a huge risotto fan but hold the dairy please as well as the traditional complexities of cooking it! For years I have been experimenting with ways to make this Italian specialty with ease and without too many heavy ingredients. So here is my version that can appeal to the vegans as well as those who want a little parmesan. Once you master this dish, you can easily get creative with the vegetables you add in (i.e. try asparagus and tomato at some point). Enjoy over the course of a few days or freeze for up to 2-3 months.

INGREDIENTS

¼ cup extra virgin olive oil
1 medium onion, small dice (about 1 ½ cup)
2 cloves garlic, minced
½ cup sundried tomatoes, rough chop (about 10 halves)
1 medium sweet potato with skin, small dice (about 2 cups)
2 tablespoons sage, rough chop
1 teaspoon rosemary, picked and rough chopped
1 tablespoon grated parmesan or nutritional yeast
1½ teaspoon salt
1 cup arborio rice
½ cup red wine (optional)
4½ cups water (if not using wine, add additional ½ cup water)
fresh ground pepper, to taste
grated parmesan cheese, optional

METHOD

1. Prep garlic, onion, sundried tomato, and sweet potato.
2. In a medium pot, combine oil, garlic, and onion. Sauté on medium heat until golden, about 3 minutes.
3. Add sundried tomatoes, sweet potatoes, sage, rosemary, parmesan or nutritional yeast, and salt. Mix well and cook for another 5 minutes.
4. Add rice, mix well, and then add 2 cups of liquid (wine and water). Be sure to stir frequently until liquid absorbed. Repeat with another 2 cups of liquid, then a final 1 cup stirring frequently until liquid is absorbed. Turn off heat.
5. Finish with fresh ground black pepper and more cheese (or nutritional yeast) to taste.

DEEP-DISH GLUTEN-FREE VEGGIE PIZZA

SERVES 4–6

The journey to create this recipe began with my older son, who is both gluten- and dairy-free. He loves pizza and is often found in my kitchen throwing together his own pizza crust. I asked him to create a recipe for a class I teach every Wednesday. This crust is a version of his recipe. It's deep dish-ish and slightly dense with a crispy exterior and soft interior. Smothered in vegetables and cheese if you please, it's a meal unto itself. What you don't eat, please freeze for up to 3 months!

INGREDIENTS

2½ cups gluten free all-purpose flour, King Arthur or Bob's Red Mill preferred
½ cup tapioca flour
2 packets active dry yeast (about 2 tablespoons)
1½ teaspoon sugar
1 teaspoon salt
1 teaspoon oregano, dried
½ teaspoon garlic powder
¼ teaspoon ground pepper, white or black
⅓ cup extra virgin olive oil
1½ cups very warm water

FOR TOPPING

2 tablespoons extra virgin olive oil
1 cup onions, sliced
½ cup broccoli, cut into small florets
½ cup green or red bell peppers
1 cup baby spinach
¼ cup sundried tomatoes
1½ cups tomato sauce, your favorite jar
1 cup cheese, or dairy-free version

METHOD

1. Preheat oven to 325°.
2. In a large bowl, combine all dry ingredients—flours, yeast, sugar, salt, oregano, garlic powder, and pepper. Mix well.
3. Add olive oil and water to flour mixture and gently mix with hands. Dough will be sticky. Form into a ball and let it rest in bowl, covered, for 1 hour or more. Dough will increase in size.
4. Cover a rectangular baking sheet with parchment paper and, using a rolling pin dusted with flour, gently roll dough across entire pan until about ⅛" thick all around. Pre- bake in oven for ONLY 5 minutes. Raise oven temperature to 375°.
5. Prep onions, broccoli, peppers, spinach and sundried tomatoes. Heat olive oil in large pan and sauté on medium until tender (about 3 minutes).
6. To make pizza, add your favorite tomato sauce to the prebaked crust, then top with sautéed vegetables and cheese of your choice. Bake for 10 minutes or until crust is golden and cheese is melted.

BUTTERNUT SQUASH TURKEY CHILI

SERVES 4–6

This dish is so simple to make yet screams with flavor. Chili is a valuable staple for any home. We serve it straight up on a bed of wilted spinach, with mashed avocado and tortilla chips, over rice, in a taco shell or wrapped up with other fixins. In other words, lots of versatility here. If you can't find a ground turkey that makes you comfy (organic pasture raised can be hard to find), opt into ground chicken (Murray's is a widely available and adored brand) or for vegetarians and vegans, ground tempeh is a good swap. Also, what you don't eat today can be eaten over the next two days or frozen for a feast down the road.

INGREDIENTS

3 tablespoons extra virgin olive oil
1 yellow onion, diced
3 cloves garlic, chopped
1 tablespoon ground cumin
1 tablespoon chili powder, medium spice
1 teaspoon salt
1 lb. ground turkey, dark meat preferred
1 cup yellow pepper, small dice
(about ½ large pepper)
1 cup butternut squash, medium dice
(about ¼ of small squash)
15 ounce can white beans (cannellini or other), drained and rinsed
28-ounce can diced tomatoes
1 cup cilantro, rough chopped

METHOD

1. Prep onion, garlic, pepper, squash, and cilantro.
2. Then in a large pot, heat oil on medium and sauté onion and garlic until tender, about 3 minutes. Add cumin, chili powder, and salt; mix well.
3. Add ground turkey to pot and break meat apart while cooking so it lightly browns; cook for about 5–7 minutes.
4. Add pepper, squash, beans, tomatoes, and cilantro and mix well. Cover pot and let cook, on medium low, for 20 timed minutes. Please be sure to check chili and stir during the cooking time. Add more salt if needed.

RATATOUILLE-SMOTHERED WILD SALMON

SERVES 4–6

This dish is a meal unto itself plus simple to prepare, nutritious, and delicious. But let's talk salmon! Without getting into too much detail, please just go wild—it's healthier for you and our environment. But I will offer this: there is no such thing as organic salmon, and much of the salmon you see in your local market is farm raised (often a light pink versus the deep pink of its wild counterpart). Buying US-caught fresh frozen wild salmon is a perfect solution (FYI, most of the fish you see in a market has been thawed at least once). Vital Choice is my go-to source, and since salmon is an edible must for me, I order a monthly box. If cooking for one, halve this dish and save the second portion for the next day's lunch. Also, try the ratatouille over another fish. Or even atop spaghetti squash, pasta, or poultry.

INGREDIENTS

1½ lbs. wild salmon fillet
1 lemon, for cleaning fish
1 teaspoon salt, for cleaning fish

FOR RATATOUILLE SMOTHER
4 tablespoons extra virgin olive oil
2 cups thinly sliced red onion (about 1 large onion)
2 cloves garlic, minced
1 tablespoon dried oregano, or fresh picked
1 tablespoon dried thyme, or fresh picked
1 teaspoon salt
2 cups grape tomatoes, halved
1 cup diced yellow pepper (about 1 small pepper)
1 tablespoon jalapeno pepper, minced (about ½ medium pepper)
2 cups baby spinach (packed)
½ cup basil leaves, loosely packed

METHOD

1. Preheat oven to 350°.
2. To clean fish, wash with salt and lemon juice, rinse under cold water, and pat dry with a paper towel. Place in baking dish.
3. Heat oil on medium low in a large sauté pan. Add onions and garlic and sauté until lightly browned.
4. Add dried oregano, thyme, salt, tomatoes, yellow pepper, and jalapeno and sauté for 5 minutes. Then add spinach and basil. Gently toss until greens are wilted.
5. Once vegetables are done, smother fish with mixture and bake in oven for 20 minutes (well done salmon). If you like your salmon medium, bake for 10 minutes only.

DRUNKEN SLOW-ROASTED CHICKEN

SERVES 4–6

When I cook, packing a dish with as much nutrition as possible is my thing. That can be a challenge when you have one child who is a picky eater and a husband with a limited palette, but it never stops me! There are so many edible celebrities in this medley, and the flavors are truly spectacular. But since chicken is truly at the center of this dish, let me offer a little guidance on how to choose. I use the same rules for eggs—organic pastured with a high welfare seal of approval. If not, opting into just organic is OK. I also frequent a small local poultry farm nearby, as knowing my farmer adds immense value! If cooking for one, you can either halve the dish or eat what you can and save the rest for meals day two and day three. Also, do note that pairing with a grain like quinoa or rice adds some serious yum.

INGREDIENTS

4 chicken breasts on the bone
1 lemon, for cleaning chicken
1 teaspoon salt, for cleaning chicken
¼ cup extra virgin olive oil
8 cloves garlic, rough chop
2 large yellow onion, thinly sliced
1 tablespoon fresh ginger, grated or minced
1 teaspoon dried turmeric
2 cups grape tomatoes, halved
½ cup broth
1 cup red wine (optional) or broth
1 cup fresh cranberries
1 cup pine nuts
1 tablespoon fresh rosemary, rough chop
1 tablespoon fresh thyme, picked
Salt, to taste

METHOD

1. Preheat oven to 325°; make sure that the oven rack is centered in the oven (not too low and not too high).
2. Clean the chicken with salt and lemon juice, rinse under cold water, and pat dry with a paper towel. Set aside.
3. In a large, thick-bottomed covered pot or Dutch oven, heat olive oil on medium and add garlic, onions, and ginger. Sauté for about 3–5 minutes until translucent. Add turmeric and tomatoes and sauté for another 3–5 minutes.
4. Add chicken breasts to pot, meat side down, and brown for 3–5 minutes, then flip over. Add the broth, wine, cranberries, pine nuts, rosemary, and thyme.
5. Then cover pot, remove from stove, and place in oven. Cook for one hour, then turn oven off and let pot sit in warm oven for another 15 minutes.
6. Salt to taste.

I feel hopeful that with the Edible Resources, 7-Day Meal & Snack Map, Ingredient Map, Essential Food Shopping Guide, and Recipes you now have enough tools to boost your Calcium ATPase while also jump-starting a fairly comprehensive health-supportive edible regime. This way of eating is really a way of life versus a diet plan. If you are new to these ideas, know that transition can be hard. But once you make it over that "change threshold," going back to what was will be even harder. Trust me! I am sorry that I can't be there to hold your hand, but stay connected at BrundeBroady.com and you will get plenty of love.

Brunde's next two chapters dive into exercise and stress reduction, both of which are essential parts of my everyday. Add the following chapters to the Toxic Reduction Plan (Chapter 18) to this, the Flexible Food and Nutrition Plan, and you pretty much have an actionable package to boost Calcium ATPase and your overall health.

CHAPTER TWENTY-ONE

CAMP TARGETED EXERCISE PLAN

PRINCIPLE ONE	Aim for 30 Minutes of High-intensity Interval Training 3–5 Times per Week
PRINCIPLE TWO	Incorporate Structured Strength Training 3 Times per Week
PRINCIPLE THREE	Embrace the Outdoors

I am neither a certified physical trainer nor any kind of fitness expert, but I do feel it is important for me to share what I have learned about Calcium ATPase and exercise. What I hope to contribute to the discussion is the fact that regular exercise can increase Calcium ATPase levels both in your skeletal muscle and heart. These are worthy goals for your future and current health. Even better news: you can increase Calcium ATPase in your muscles whatever your age. First off, if you are new to exercise or have any health concerns, please check with your doctor before following my recommendations. My guess is that this plan likely aligns with what you are currently doing. And if so, cheers! That makes my job easy.

PRINCIPLE ONE
Aim for 30 Minutes of High-Intensity Interval Training 3–5 Times per Week

The Science

One study examined the effect that eight weeks of interval treadmill training had on Calcium ATPase levels. Forty-two mice were divided into two groups: trained mice and their more sedentary brethren. The trained mice ran on the treadmill at a 10% incline, working up to four intense bouts of exercise of 2.5 minutes each. These bouts were each separated by a 3-minute rest period. The mice were trained five days a week. The control group mice had no exercise added to their normal lives. After the eight-week training protocol, Calcium ATPase levels increased by 25% in the trained mice group versus the untrained mice control group.[384]

Another group of researchers looked at the impact that both moderate and high-intensity exercise had on Calcium ATPase RNA (the gene precursor to Calcium ATPase). Fifty-two rats were divided into three groups: sedentary rats, moderate-intensity trained rats (treadmill for sixty minutes, five times a week at 0% incline), and high-intensity trained rats (treadmill for five one-minute sprints, five times a week at 15% incline). All trained rats followed six-week programs.

Muscle biopsies of the gastrocnemius (the big muscle in the calf in humans) were taken from each of the rats before and after training. Results showed that rats in both exercise protocols had significantly increased Calcium ATPase RNA over the sedentary control group. Moderate exercise resulted in an 83% increase, whereas high-intensity exercise resulted in over a 100% increase.[385]

In addition to stimulating Calcium ATPase in the skeletal muscle, exercise training has been shown to increase Calcium ATPase levels in cardiac tissue. Researchers genetically modified mice so their heart condition approximated heart failure in humans. These mice then completed a six-week, five-day-a-week training protocol that consisted of a 20-minute warm-up, followed by

384 Ferreira J,Bacurau A, Bueno C, Cunha T, Tanaka L, Jardim M, Ramires P, Brum P. Aerobic exercise training improves Ca2+ handling and redox status of skeletal muscle in mice. Experimental Biology and Medicine 2010; 235:497-505.

385 Kubo H, Libonati JR, Kendrick ZV, Paolone A, Gaughan JP, Houser SR. Differential effects of exercise training on skeletal muscle SERCA gene expression. Med Sci Sports Exerc. 2003 Jan;35(1):27-31. doi: 10.1097/00005768-200301000-00005. PMID: 12544631.

ten four minute intervals at 85% Vo2 maximum, alternated with 2-minute active recovery, for a total of 1.33 hours of exercise. At the end, they showed an increase of 25% in Calcium ATPase levels in the heart, which resulted in significant improvements in cardiac function.[386]

A similar study looked at the same training protocol in rats after an induced heart attack. Post-heart attack, Calcium ATPase levels were approximately 30% lower as compared to normal rats. After completing the training protocol, the heart attack group's Calcium ATPase levels returned to normal. In addition, normal rats that followed the exercise regimen had 82% higher cardiac Calcium ATPase levels compared to sedentary rats.[387]

The Action: In Motion

The great thing about high-intensity interval training is that you can do it with the exercise of your choice. Some easy options are to program an exercise bike, treadmill, elliptical, or row machine to simulate interval training by increasing the exercise intensity automatically. Another option is simply to increase the speed and/or intensity based on your own perceived exertion.

I personally like a programmed setup on the treadmill, but I also take online classes from time to time. If you need inspiration, Google high-intensity interval training, and you will find many options!

PRINCIPLE TWO
Incorporate Structured Strength Training 3 Times per Week

The Science

Aging Muscle and Calcium ATPase

As in other parts of the body, including the heart and brain, Calcium ATPase levels decline in our muscles as we age. The net result is reduced muscle strength and endurance, which is commonly

386 Høydl M, Stølen TO, Kettlewell S, Maier L, Brown J, Sowa T, Catalucci D, Condorelli G, Kemi O, Smith G, Wisløff U. Exercise training reverses myocardial dysfunction induced by CaMKIIδC overexpression by restoring Ca2+ homeostasis. J Appl Physiol (1985). 2016 Jul 1;121(1):212-20

387 Wisloff U, Loennechen J, Currie S, Smith G, Ellingsen O. Aerobic exercise reduces cardiomyocyte hypertrophy and increases contractibility, Ca2+ sensitivity and SERCA-2 in rat after myocardial infarction. Cardiovascular Research 54 (2002) 162-174.

chalked up to getting old. Although it is bad news that Calcium ATPase levels in our muscles decline as we age, the good news is that exercise can increase Calcium ATPase even in aging muscle. Let's look at some research.

One study compared Calcium ATPase levels in the quadriceps muscles of seventeen women aged eighteen through thirty-two to those of eleven women aged sixty-four through seventy-nine. The study found that the older group had 38% lower levels of Calcium ATPase compared to the younger group. Calcium ATPase levels were again measured after a twelve-week, high-resistance training regime, consisting of three sets of eight repetitions of maximum weight performed three times a week. After the resistance training, Calcium ATPase levels increased in elderly women by 23% as compared to pretraining levels.[388]

Another study examined age-related changes in Calcium ATPase levels in the quadriceps muscle of elderly vs. young men. Calcium ATPase levels in elderly men, aged sixty-eight through seventy, were found to be 35% lower as compared to young men, aged twenty-eight. In addition, the study compared nonexercising elderly men to elderly men who followed a strength training regimen. Calcium ATPase levels were 31% higher in the strength-trained group as compared to the sedentary elderly men.[389]

The Action: In Motion

The specific exercise protocol in the studies referenced above consisted of 3 sets of 8 reps at the maximum weight possible performed 3 times per week. Given the research, I would recommend that as a first choice, but it is likely that strength training in whatever form has a positive impact on Calcium ATPase. One great thing about the Internet and home fitness equipment is that there is no shortage of strength training options. Find the one the works for you.

Remember: strength training will take time to show results. Immediately after an intense workout, your Calcium ATPase levels will likely take a dip making your feel depleted and sore at the

388 Hunter S, Thompson M, Ruell P, Harmer A, Thom J, Gwinn T, Adams R. Human skeletal sarcoplasmic reticulum Ca2+ uptake and muscle function with aging and strength training. Journal Applied Physiology 1999.
389 Klitgaard H, Ausoni S, Damiani E. Sarcoplasmic reticulum of human skeletal muscle: age-related changes and effect of training. Acta Physiol Scan. 1989,137,23-31.

same time. Have faith: over time you will have a higher baseline of Calcium ATPase to power you through your workouts!

PRINCIPLE THREE
Embrace the Outdoors

Although there are no research studies to back up the positive effect of spending time outdoors on Calcium ATPase, I do feel confident in saying this: if being outdoors makes you happy, and you enjoy activities such as walking, hiking, surfing, swimming, or biking, then try to add these in when possible. Not only will you gain the physical benefit of activity, but also the stress reduction component that will support your Calcium ATPase levels. My favorite activities are taking nature walks on my parent's ranch in Texas (the few times a year I can visit, with my dad still leading the charge), and faster-paced walks with friends in Central Park. Sometimes I can even drag Knute away from the computer to take a long walk when we are in the country. Pay attention: if any outdoor activity clicks with you, enjoy!

CHAPTER TWENTY-TWO

CAMP STRESS-REDUCTION PLAN

THIS PLAN HAS ONLY ONE BASIC PRINCIPLE!	Embrace Mindfulness Practice Daily

I hope this chapter does not sound cliché when I ask you to embrace a mindfulness practice. Likely you have been hearing this from every book you have read on health, from your doctor, from your friends, and even now from apps that pop up on your phone. Please hear me out. The reason why I have included it in this book is specific. As we learned in the chapter on stress, stress hormones, in particular catecholamines, have a negative impact on Calcium ATPase. Most of the research that has been done focuses on formal practices such as yoga and meditation and qigong, all of which are more structured and can require a significant time commitment.

However, stress reduction and hence catecholamine reduction can occur during various activities. The key is to find what works for you. I will give you more suggestions below, but first let's look at some research detailing the effect yoga, meditation, and qigong have on catecholamines.

The Science

Research has demonstrated that the practice of yoga, meditation, and qigong can reduce catecholamine levels. One study compared the catecholamine levels of a group of nineteen subjects, eleven men and eight women between the ages of eighteen and forty, who had practiced transcendental meditation (TM) for at least twelve months, twice a day. The control group consisted of sixteen healthy subjects ranging in age from twenty-two to thirty-five who did not perform any relaxation techniques. Blood samples of catecholamine levels of both norepinephrine and epinephrine were significantly lower in the TM group, with norepinephrine levels an astonishing 42% lower as compared to the control group.

In addition to TM, numerous studies have demonstrated that a regular yoga routine reduces the levels of epinephrine, a key catecholamine. For example, one study measured the effects of a yoga routine consisting of one ninety-minute session once a week with an instructor, supplemented by daily home practice of forty minutes (utilizing a DVD) for twelve weeks. The test group consisted of twelve yoga-trained individuals, with thirteen non-yoga-trained individuals serving as the control group. Stress hormone levels were measured before and after twelve weeks. At the beginning of the study, stress hormones were the same in both groups. After twelve weeks of yoga, the training group's stress levels had declined by 40%. In addition, the antioxidant level for glutathione was increased by 200% in the yoga group, which benefits overall stress reduction by significantly reducing oxidative stress. Oxidative stress has a negative effect on Calcium ATPase function.

Another study consisted of eighteen participants who engaged in yoga sessions once a week over a period of four months. They were also given yoga home assignments. As compared to the beginning of the program, norepinephrine levels in the yoga-treated group declined approximately 30% over four months.

Researchers also examined the effect of a three-month yoga regime on epinephrine in three age groups: twenty to twenty-nine years, thirty to thirty-nine years, and forty to fifty years. Epinephrine levels declined by 60%, 30%, and 42, respectively.

For those who enjoy higher activity levels, more vigorous practices of the ancient martial arts of Tae Kwon Do and qigong have been proven to lower catecholamine levels.

One study consisted of twenty women ranging in age from sixty-six to seventy-four. Ten undertook Tae Kwon Do training of sixty minutes per day, three days per week, for a period of twelve weeks, while the other ten women served as the control group. Not only did the active group have significant reductions in catecholamine levels, they also had reductions in arterial stiffness and blood pressure and increases in skeletal muscle strength.

Similarly, in a clinical trial of fifty-eight patients, twenty-nine undertook a ten-week training in qigong, and the other twenty-nine patients were utilized as the control. In the qigong group, there were significant reductions in norepinephrine, epinephrine, systolic blood pressure, diastolic blood pressure, and other stress indicators.

The bottom line is regularly calming our minds has scientifically quantifiable benefits to our health and wellbeing. Next, let's take a deep dive into the various iterations of yoga and meditation.

The Action: Mindfulness Practices

These studies are a bit intimidating. Meditation twice a day every day. Yoga every day for 45 minutes. Don't be discouraged, I encourage you to start somewhere, wherever that is. Remember stress reduction is one part of the plan you are following to maximize your Calcium ATPase so perfection is not required in every component all of the time. As someone who has been involved in a mindfulness practice for over 30 years what I can share from my experience is this: My life is better when I am incorporating mindfulness into my life on a daily basis then when I stop. It is that simple. Today there are so many apps, so many books, so many videos, you have so many options. But if you are starting try not to make it too complicated. At the end of the day all that you need is yourself and your breath.

If an overview of yoga, meditation practices, and qigong would be helpful, I am including this below for your reference or perspective, or both!

YOGA

Yoga is a broad and multifaceted practice, and for that reason (and your benefit), we're going to delve a bit deeper into the subject. For increasing numbers of people, yoga is becoming the go-to for reducing stress and increasing overall health. Not only do we know that reducing stress improves our Calcium ATPase levels, but scientific study after study shows how much yoga helps in many other areas of life, including providing an overall sense of well-being and happiness. How could we not want more of that?

For those of you who already have an established yoga practice, well done. Your Calcium ATPase thanks you! For those who would like to have a yoga practice and don't know where to begin, read on.

First, you don't have to be flexible to do yoga. Many people mistakenly believe it's for bendy people only. But anyone can take it up, and over time you may experience an increase in your comfortable range of motion—or not. In any case, flexibility is not a requirement to begin. You can have two replaced knees and an achy hip and get enormous benefits from doing regular restorative yoga. If you have any medical concerns, ask your doctor. That is, if they haven't already recommended you give yoga a try.

Remember to start gently and increase your intensity over time (like with most things). The key is to do yoga regularly so the beneficial effects can accumulate gently over time. Make a note of what your first class felt like, then after a month, and then a year. You might be astonished at how much better your body feels and how much more attuned you are to its needs. (And, yes, you might be more flexible, too.)

There are many different types of yoga to choose from. Explore them. Find the style and teacher that suits you best. Below we've outlined a few of the most popular types, and please remember that the descriptions may vary from studio to studio.

Hatha

This is often used as a blanket term for any yoga that involves postures (asanas) and breathing exercises (pranayama). So if you find a studio offering hatha yoga, be sure and ask what their specific style means. In general, it will be a balanced approach.

Vinyasa

Sometimes called *vinyasa flow*, this is a quite popular movement-based yoga. The teacher guides her students through a series of postures designed to open and work the entire body, although an emphasis can be placed on specific areas, such as the upper or lower body. Everyone does the same poses, and the teacher gives detailed point-by-point instructions on technique and transitions. Vinyasa comes in many levels of intensity, so be sure and check that out. Your studio might offer Gentle Vinyasa, to an "All Levels" class, which is theoretically for anyone but might cater to those more advanced in their practices.

Iyengar

B.K.S. Iyengar, a yoga master from India, developed this lineage. It's distinct from vinyasa in that the poses are static, held for longer, and have strict rules for alignment. Props are used to help students maintain the poses comfortably. Overall, Iyengar is considered to be a less intense and more therapeutic practice.

Ashtanga

A separate lineage from Iyengar, and considered the foundation for vinyasa, ashtanga yoga is an intense and disciplined practice. The primary series, which all students must master to move on to the intermediate series, consists of the same forty-one poses, each counted to a specific deep breath. The students most often practice in *mysore* style, meaning they show up to the studio between a set number of hours and practice on their own. Everyone in the studio at a certain time will do the same series of poses, but at their own pace and timing. A teacher walks among them giving individual adjustments and guidance. The teacher also determines when a student is ready to "receive" another pose or move on to the next level of practice. While ashtanga is for anyone and good studios have all levels of students, this is a more intense and deep practice. Some ardent followers travel to India to study with the masters themselves.

Hot or Bikram

Although mired in controversy, bikram yoga and its more benign namesake, *hot yoga*, all share a common theme: the yoga is performed in a room heated to 105F with 40% humidity, and students sweat in streams. Bikram yoga students do the same twenty-six postures at the instruction of a teacher who has memorized a preestablished script. While the poses are general and strengthen the entire body, the heat is the kicker and requires stamina and strength. Hot yoga is open for interpretation by the studio and individual instructor. They might do a modified version of bikram or do some other vinyasa flow-type class, but in the same style of heated room. Be sure and check out the specifics of each class, as they vary. But whether bikram or hot yoga, these classes will be intense and push you to your limits, hence their popularity.

Yin and Restorative

At the quiet end of the yoga spectrum rest these two lovelies. Yin yoga entails deep, carefully aligned stretches, held for several minutes. One of the core principles is stretching connective tissue, such as fascia, ligaments, and tendons, in a controlled and healthy manner. Classes are quiet and dimly lit, and deep breathing is encouraged. Restorative yoga can include yin yoga, but it can also include props and poses modified to help out those without much flexibility due to injury, lack of muscle use, or just plain genetics. These classes can be deeply relaxing, and you can almost hear the Calcium ATPase getting healthier as your stress fades away.

Many other forms of yoga exist, and if they appeal to you, check them out. There's aerial yoga, where you swing from the ceiling on soft parachute-like material; kundalini yoga, where you waken dormant energies within you with fast breathing; and trim-n-tone yoga, where you use light, hand-held weights during a vinyasa class. Online yoga is popular with all of the above options and is becoming especially important during the Covid-19 pandemic. Find what's right for you and keep up a regular practice. You and your Calcium ATPase will be grateful.

Meditation

Millennia of experience and positive results from meditation must mean something. Finally, science is catching up and proving through numerous studies that a regular meditation practice reduces stress (hello, healthy Calcium ATPase) and increases a sense of well-being and happiness, leading to a healthier mind and body. If you have a meditation practice, then you already know this. But if you don't, you may wonder how to get started, or if you're even capable of meditating. How many times have you heard or thought, "I can't meditate because I can't get my mind to stop."

Getting your mind to "stop" isn't necessary for meditation. Simply being aware of the chatter and letting it go is enough. Below, we discuss several types of meditation techniques and how to go about starting. Your community may have meditation classes at yoga studios or religious centers such as churches, mosques, or synagogues, or perhaps even a Buddhist center. Online courses and resources abound. Check them out and see what might suit your particular style.

Guided

This form of meditation entails listening to a teacher, either live or via an app, talk you through a sequentially deeper state of meditation and relaxation. They will have a process wherein you begin to calm, develop a more fine-tuned sense of awareness, and then achieve deeper states of relaxation. The good part about this form of meditation is that the listener just sits back and relaxes, letting him- or herself be taken on a peaceful journey. Some schools of meditation have a set visualization that practitioners use every time; others use a variation.

Mindfulness

Much ado has been made about mindfulness in recent years, and with good reason. Slowing down and becoming increasingly aware of ourselves, our lives, and every single entity within it help build gratitude and calmness. You can find plenty of books, online courses, meditation centers, and retreats that offer deep dives into the practice of mindfulness. In short, mindfulness is maintaining awareness of everything around and within us at any given moment. Place your feet on the ground. What does that feel like? What does it bring to mind?

Mantras

Some meditation practices use mantras to achieve deeper levels of peace and awareness. A mantra is a word, phrase, or sound you repeat over and over. The classic example is the word *Ohm*. This simple mantra can be said out loud or within the privacy of one's own mind to great meditative effect. Other forms of meditation, such as *Transcendental Meditation (TM)*, have teachers trained to levels such that they give the student their mantra in a private ceremony.

Qigong

This long-established exercise and stress-management system may be one of the least-known practices, but it is growing in popularity and is worth exploring as you seek enriching ways to nurture your body's ability to increase and optimize Calcium ATPase levels.

Let's begin unraveling the mystery of what it is and why we've included it in our Program section, with some pronunciation help and a working definition. Qi, pronounced "chee," is the term for life force or life energy and gong, pronounced as "gong," means skill that is developed through consistent work or practice. Taken together, a good way to think about this term is a practice or set of exercises designed to cultivate more life force or energy in the body to improve overall health. You have likely seen groups of people in parks flowing together in a series of movement progressions and wondered if you could learn to do something so graceful and beautiful. Not only can you, but we encourage you to do so.

These exercises are both beautiful and functional and have been shown to improve circulation of blood and lymph; increase ease of movement and range of motion by supporting fascia, ligaments, and tendons; support the cardiovascular and respiratory systems; improve immune function; lower stress and anxiety levels; improve sleep quality; and improve cognitive functioning, all of which can enhance Calcium ATPase levels. Qigong is an ancient practice that is ideally suited to meet the modern demands of an always-on society. From two minutes of one exercise sandwiched between clients to a more formal one-hour comprehensive class, this infinitely adaptable health art may be just what the doctor ordered (and many do).

Like yoga and meditation, there are many variations and forms, and a quick search of the Internet and YouTube can be a little daunting, so we'll describe a few classic exercise sets and offer some advice on how to find a good teacher for in-person training.

Eight Pieces of Brocade

This is one of the more ancient forms in qigong. The eight exercises can be practiced standing or sitting and can therefore be undertaken by most people regardless of current fitness level. The eight sections, each named to evoke the kinds of movements undertaken, are designed to address specific regions of the body and as a whole, work together to improve overall health and function.

Daoist 5 Yin Tonification Set

In Classic Oriental Medicine, there are five Yin organs in the body—the Lungs, the Kidneys, the Liver, the Heart, and the Spleen—and this classic exercise set includes a series of movements designed to improve the energetic flow in the associated meridians. They can be practiced individually or as the set.

Five Animals Frolic

This ancient exercise set was designed to emulate the movements of different animals—the Crane, the Deer, the Tiger, the Monkey, and the Bear—and like the Taoist 5 Yin set is associated with the Yin organs. It is a fun set to practice and can be an engaging way to include children.

Six Healing Sounds

These sounds, again associated with specific organs and meridians, are said to cleanse stagnation from the associated meridian in order to improve health. They can be performed individually or as a set.

Some general thoughts: most of these exercises can be safely learned, at least at a rudimentary level, from a good video. Our recommendation is to find a good teacher in your area and learn the exercises in person whenever possible. These instructors will often have made their own videos and can facilitate your learning with a combination of in-person and video practice. They can also guide you to commercially available videos they prefer for continuity of learning the underlying fundamentals.

Two good resources for finding a qualified instructor in your area:

- The National Qigong Association: www.nqa.org, which has a "Find a Teacher" tab on the main page of its website.
- Qi—The Journal of Eastern Health and Fitness: www.qi-journal.com, which has a "Professional Listings" tab on their main page as well as in the print version.

CHAPTER TWENTY-THREE

FUTURE DIRECTIONS

In this book, you have been made aware of the importance of intracellular calcium in health and disease. Simply put, when calcium levels within our cells are not properly regulated, health problems occur. The flip side is that when calcium regulation is maintained, we are less vulnerable to many chronic diseases, such as cancer, diabetes, heart disease, and Alzheimer's.

The good news is that the interest in intracellular calcium and Calcium ATPase as a key regulator of cellular health has been growing over the past 20+ years. A second wave of research is now occurring regarding the search for compounds and gene therapy to stimulate Calcium ATPase as a means of treating diseases. As I have highlighted throughout the book, significant progress has been made in the areas of obesity, diabetes, Alzheimer's, heart failure, and muscle disease.

Given its importance, you may be asking yourself, *How can I get my Calcium ATPase levels tested?* The answer is that currently there are no tests available outside of research studies. I am trying to change that!

In the meantime, we have learned that A1C levels are negatively correlated with Calcium ATPase. If your A1C levels are high, your Calcium ATPase levels are likely low. Obesity is also correlated with low Calcium ATPase.

Another common test done for LDL cholesterol may offer clues. LDL inhibits Calcium ATPase levels in your heart, so if you have high LDL, your Calcium ATPase may be less than optimal. The same is true for sleep apnea.

Someday soon, I hope you will be able to get your Calcium ATPase levels tested easily, but until then, I hope I have given you knowledge and tools to manage this important part of your health.